Praise for Gerald M. Stern's

THE BUFFALO CREEK DISASTER

"[A story] well worth telling. . . . This book could be far more valuable to society than his stunning legal victory."

—*The New Yorker*

"I could not put it down."

—Anthony Lewis, author of *Gideon's Trumpet*

"Exciting. . . . The reader cheers Stern on, as much to participate in his response to a cry for help as in his victory."

—*BusinessWeek*

"The importance and the drama of this account lie in the skilled gathering of details that become, fittingly, a tidal wave of facts to wash away Pittston's defense."

—*The Washington Post*

"Totally absorbing—one of the best reads of the year."

—*The Washington Monthly*

"Painstaking detective work . . . detailed so masterfully. . . . Readers caught in this book's spell will find themselves, despite their efforts, returning to that sleepy Saturday."

—*Newsday*

"Borders on the miraculous." —*The Cleveland Press*

"Gerald Stern is a lawyer with a story to tell. He tells it enthrallingly." —*Houston Chronicle*

"Stern's account of the furious jockeying, the stalling tactics, the horse trading and back-room dealing that went into the settlement is fascinating. Best of all are the sections that show how big-time lawyers think." —*The Baltimore Sun*

GERALD M. STERN

THE BUFFALO CREEK DISASTER

Gerald M. Stern was a founding partner of the Washington, D.C., Rogovin, Stern & Huge law firm. Prior to that he was a partner with Arnold & Porter for eleven years, where he was the lead counsel for the survivors of the Buffalo Creek disaster. Before joining Arnold & Porter he was a trial attorney with the Civil Rights Division of the United States Department of Justice, trying voting discrimination cases in the South. He wrote about those experiences in chapters of two books, *Southern Justice* and *Outside the Law*. He is also the author of the forthcoming book, *The Scotia Widows*, the story of his representation of the women who lost their husbands in the Scotia Mine disaster. He has also served as General Counsel of Occidental Petroleum Corporation and as Special Counsel to the United States Department of Justice. Presently, he is a legal consultant and lives in Washington, D.C.

THE BUFFALO CREEK DISASTER

THE BUFFALO CREEK DISASTER

How the Survivors of One of the Worst Disasters in Coal-mining History Brought Suit Against the Coal Company—and Won

GERALD M. STERN

Vintage Books
A Division of Random House, Inc.
New York

SECOND VINTAGE BOOKS EDITION, MAY 2008

 Published in the United States by Vintage Books, a division of Random House, Inc., New York, and in Canada by Random House of Canada Limited, Toronto. Originally published in hardcover as *The Buffalo Creek Disaster: The Story of the Survivors' Unprecedented Lawsuit* in slightly different form in the United States by Random House, Inc., in 1976.

Grateful acknowledgment is made to Cherry Lane Music Company and Alfred Publishing Co. Inc. for permission to reprint four lines from "Take Me Home, Country Roads" words and music by John Denver, Bill Danoff, and Taffy Nivert, copyright © 1971 and copyright renewed 1999 by Cherry Lane Music Publishing Co. Inc. (ASCAP), Dimensional Music of 1091 (ASCAP), Dreamworks Songs, Anna Kate Deutschendorf, Zachary Deutschendorf, Jesse Bell Denver for the USA. All rights for Dimensional Music of 1091, Anna Kate Deutschendorf and Zachary Deutschendorf are administered by Cherry Lane Music Company, Inc. (ASCAP). All rights for Jesse Bell Denver are administered by WB Music Corp. (ASCAP). All rights for the World excluding the USA are administered by Cherry Lane Music Company, Inc. International copyright secured. All rights reserved. Reprinted by permission of Cherry Lane Music Company, Inc. and Alfred Publishing Co.

The Library of Congress has cataloged
the Random House edition as follows:
Stern, Gerald, 1937–
The Buffalo Creek disaster.
1. Prince, Dennis. 2. Pittston Company.
3. Buffalo Creek, W. Va.—Flood, 1972. I. Title.
KF228.P75S7
345'.73'026
75—40554

Vintage ISBN: 978-0-307-38849-0

www.vintagebooks.com

Printed in the United States of America
20 19 18 17 16

For my wife, Linda, my sons, Eric and Jesse,

my daughter, Maia, my parents, Lloyd and Fannye,

and the people of the Buffalo Creek Valley

Contents

PART TWO

PART THREE

Foreword

This is the story of how a small West Virginia community responded to a horrendous coal-mining disaster that killed over 125 people and left thousands homeless. It is powerful, troubling, and uplifting. We see the humanity of hard-working coal miners and their families who worked together to fight back and hold legally accountable those who recklessly endangered their lives and devastated the valley they called home.

The Buffalo Creek drama takes place in the border area between West Virginia and Kentucky, an area known to history as the home of the infamous Hatfield-McCoy feud, a cycle of violence in search of revenge. But in this modern tale, the goal was not revenge but justice, and fair-minded legal arguments replaced guns as the weapons of choice.

Jerry Stern's account has long been required reading for students at many law schools throughout our country. He vividly captures what really went on behind the scenes in the high-stakes legal battle and the fundamental questions about how to deal with man-made disasters, which underlie the case. Instead of our judicial system, should we rely on insurance payments, a workmen's compensation system, relief legislation, or volunteer organizations? Should a matter of such importance be left to the courts when the suit will be expensive and time-consuming, against a wealthy and well-defended adversary?

The Buffalo Creek Disaster shows that the legal system can work if it remains open to everyone, with practical, fair judges and good lawyers willing to take on difficult litigation on a pro bono or low fee basis. Fortunately, many lawyers in this country do just that.

This story opens our eyes to the fact that the survivors of a disaster often suffer severe mental anguish, as well as physical harm and financial loss. What the lawyers in the Buffalo Creek case called the survivors' "psychic impairment" is now officially known as post-traumatic stress disorder. From studies of the survivors of Buffalo Creek and other tragedies, we now know that serious cases of post-traumatic stress disorder often follow disasters, as well as difficult combat service, and victims need society's help to cope with them. As our veterans return from Iraq and Afghanistan, this is an important lesson to recall.

In an era of ever-increasing mass disasters, from 9/11 to Katrina, this reissue of *The Buffalo Creek Disaster* is a timely reminder that much remains to be done after such tragedies. Though media and public attention may fade, the survivors' pain endures and demands a response.

The Buffalo Creek flood was an important chapter in our nation's history. Jerry Stern's classic work provides readers with tremendous insight into the causes of the disaster, its human toll, and how our legal system can hold negligent parties accountable, help to remedy the losses, and help victims recover.

—Bill Clinton
May 2008

Prologue

One of the worst man-made disasters in this country's history began early one morning in February 1972. A coal company's massive coal-waste refuse pile, which dammed a stream in Middle Fork Hollow in the mountains of West Virginia, collapsed without warning to the people in the long, narrow Buffalo Creek Valley below. This failure unleashed over 130 million gallons of water and waste materials—stream water from recent rains as well as black coal-waste water and sludge from a coal-washing operation. This 20-to-30-foot tidal wave of rampaging water and sludge, sometimes traveling at speeds up to 30 miles per hour, devastated Buffalo Creek's sixteen small communities.

Over 125 people perished immediately. Most were women and children unable to struggle out from under the thick black water choked with crushed and splintered homes, cars, telephone poles, railroad tracks, and all manner of other debris. There were over 4,000 survivors, but their 1,000 homes were destroyed as well as most of their possessions.

There have been many coal-mining disasters. But the Buffalo Creek disaster is unique. This time it was not the strong, working male coal miners who died in the mines. That has become so commonplace in coal mining as to be expected. No, this time it was the miners' defenseless wives and children, caught, unprepared for death, in their beds one Saturday morning.

Maybe this is the reason some of the survivors responded differently this time. Instead of accepting the small settlements offered by the coal company's insurance adjustors, a few hundred of the 4,000 survivors banded together to sue the coal company, to make the company admit its responsibility for this massive destruction, to make the company pay them for all their losses, mental suffering as well as property damage, and to make sure something like this would never happen again, to them or to anyone else living at the mercy of the coal companies.

This is the story of their lawsuit, of their use of the legal system to uncover evidence of corporate irresponsibility. And, most significantly, this is the story of our legal system's ability to respond, to create new precedent, to fashion a remedy which may permit any one of us to recover for the mental suffering which normally follows when we survive another's reckless act, physically unharmed but mentally scarred.

This is also my story, the story of what this lawsuit meant to me as the survivors' lawyer.

PART ONE

I

The Call for Help

The Buffalo Creek disaster was news throughout the country back in February 1972. A massive dam used by a coal company to filter the black waste water from its coal-cleaning plant had collapsed, and a seventeen-mile valley of small coal-mining towns lay in ruins. Over 125 people died, and thousands more were left without homes or possessions. The nightly news on TV showed pictures of fast-moving water, fleeting glimpses of dead bodies covered with black water, people sobbing their stories to interviewers while the camera bore in even closer on their pained, contorted faces. Another disaster had occurred, this time a big one, one of the worst in West Virginia's long history of coal-mining disasters. And a disaster is always news.

As the news reports came in during those first weeks after the Buffalo Creek disaster, however, I went on doing what I was doing, which was nothing. It was strange that I paid so little attention to Buffalo Creek because I was wishing there was something I could do, something with purpose. I had once found it. In the early 1960s, right out of law school, I'd traveled around Mississippi, Alabama, and Louisiana trying voting discrimination cases for the government. I had a purpose then—helping blacks in the South who were slowly beginning to ask for their rights.

My going South at that particular time constituted a change in my plans, since I had already accepted a job with a Wall Street law firm and was supposed to start soon after graduation. I had figured I'd learn some corporate law, and then after a few years I'd take my legal experience home to Memphis. But the election of President Kennedy and the Freedom Rides changed all that. When I saw the Freedom Riders' Greyhound bus at Anniston, Alabama, burning grotesquely on the Huntley-Brinkley nightly news, I called Wall Street and told them I had changed my mind. Instead, I joined the Civil Rights Division of the Department of Justice in Washington.

I had to go and help. And no one seemed better qualified than I. I was white and a Southerner. And, as a lawyer, I'd have a role, an identity, in a structured system which would permit me, indeed encourage me, to rise to my feet and plead someone else's case. I'm usually too shy to ask for something for myself. But in my lawyer's suit, I'm always able to speak up for others.

Two years with the Civil Rights Division exhausted me, and when President Kennedy was killed, I couldn't go on. By then, though, the civil rights struggle was well under way. Our small group of government lawyers had mushroomed, college kids from all over the country streamed South to help, and the 1965 Voting Rights Act guaranteed the vote to black Southerners without requiring that the government prove, county by county in endless court trials, that there was discrimination.

After I left the Justice Department, I joined Arnold & Porter, a Washington, D.C., law firm. For the next nine years I played lawyer for a different breed of clients, primarily corporations fighting the government. The little purpose this job held was harder to see. Sometimes I could convince myself I was standing by the little guy. Maybe it was good to spend three years of my life helping a small railroad fight off extinction at the hands of the wealthiest railroad-banking complex in the country. But more often than not, I had to blind myself to the end in sight and find my purpose in doing the day-to-day lawyering job well.

At that time our law firm decided to permit one partner each year to spend all his time on *pro bono publico* cases—public interest cases. The third year of the program it was my turn. Halfway through that year, I found myself restless, staring out the window. I hadn't found any case to fill my days and nights, to make me feel bigger than myself. I even went to see a psychiatrist. I told him I was bored. I said I lacked purpose. He said, "That will be forty dollars, please." I didn't go back. I didn't much believe in psychiatrists then.

Soon after, one of my partners called. "I just heard from a friend with the Environmental Defense Fund. He wants to know if we'd be willing to represent some of the survivors of the Buffalo Creek disaster." I said I'd get in touch with Harry Huge, another one of my partners, to see what this was all about. I still hadn't read much of the news accounts of Buffalo Creek.

I figured Harry would know about this coal-mining disaster because he had just won a multimillion-dollar verdict for disabled coal miners and widows against the United Mine Workers Welfare and Retirement Fund. This fund had been set up years before by the coal companies and the United Mine Workers' president, John L. Lewis, to provide benefits to injured coal miners and to families of coal miners killed during work. Even to this day the Welfare Fund's checks are known as "the John L." This historic fund, the first such company-employee trust fund in the nation, was supposed to be administered by independent trustees, responsible not to the coal companies or the United Mine Workers, but to the beneficiaries of the fund. Instead, for years the Welfare Fund maintained millions of dollars in a checking account in the United Mine Workers' bank in Washington, D.C., depriving the pensioners and widows of millions of dollars of interest which could have been earned on their money if it only had been placed in savings accounts or had been invested in government securities.

Just months before the Buffalo Creek disaster, a federal court in Washington had found that the union-controlled bank, the union, the bank president, and one of the fund's trustees were liable to pay a total of $11.5 million to Harry's clients and the entire class of disabled miners and widows who were beneficiaries of the Welfare Fund. There were many such disabled miners and widows in West Virginia. Harry Huge was now a folk hero in the hills of West Virginia.

So I was not surprised when Harry told me he'd already been down to Buffalo Creek. He said the Buffalo Creek Citizens Committee had called him soon after the disaster to ask if he would represent them. This committee had been formed within two weeks after the disaster, even before all the missing and dead had been accounted for, when a group of angry survivors gathered at the Buffalo Creek Grade School. They wanted revenge.

It was not easy for them to meet publicly, and openly, to plan an attack on a coal company, especially one which was one of the largest employers in their county, Logan County, West Virginia. But the people were determined. They immediately elected a committee of two members from each of the sixteen communities in the disaster. This fair, open election was unusual for Logan County, which is known for its political corruption and stuffed ballot boxes. The last county election, the May 1970 Democratic primary, had been such a blatant fraud that some people connected with the winning faction were convicted in federal court of conspiracy to cast fictitious votes for candidates. As the

United States Supreme Court found in 1974, if a Logan County voter needed help in using the voting machines, the man working for the Hager slate, the winners, "would join the voter in the voting machine and, aligning their bodies so as to conceal what they were doing, would put the votes on the machine for the entire Hager slate" rather than for the other slate the voter wanted to vote for. "In addition, he [the Hager slate's man] simply went into the voting machine on his own and cast many fictitious ballots." The convicted conspirators, now known as the Logan County Five, were Logan's state senator, sheriff, circuit clerk, county clerk, and deputy sheriff.

So, the survivors' open election in the Buffalo Creek Grade .School was unique for Logan County, real votes being cast by real people who expected their votes to have real meaning. They chose Charlie Cowan as their chairman. Charlie runs a gas station at Amherstdale, located halfway down the Buffalo Creek Valley. Again the irony—because he is black and independent, the Logan County powers, years earlier, had denied Charlie the right to vote. Certainly they would never let him run for office. Now, however, huddled together as equal victims of the disaster, the people of the Valley, most of whom were white, chose this black man to lead them.

Charlie is fifty-six years old. He's run his gas station at Amherstdale for twenty years. That station is like the big tree on the front lawn of most county courthouses. It's the gathering place where people just hang around, talk, and get the latest news. Charlie leaves his television set on all the time. The people wander in to buy Cokes or candy or milk or little trinkets for the children and just leave the money on the cash register or fill out a little piece of paper showing how much they owe.

The people are a little awed by Charlie because he is the father of Buffalo Creek's most famous athlete, Charles Cowan, Jr., the local football hero who went on to become an offensive tackle for the Los Angeles Rams. Junior has been the Rams' right tackle for fourteen years, missing only four games in all that time. He holds the Rams' record for playing in two hundred consecutive games. Charlie and his son are quite alike—big, strong, and determined. That's one reason the people picked Charlie to lead them in their fight against the coal company.

They also picked Charlie because he'd stood up to the coal companies in the past when he brought a lawsuit against them for strip-mine damage to his property. Charlie decided, from his experience in that case and from other legal battles with the coal companies, that the first thing he would have to do for himself and for the other survivors of this disaster would be to get legal help.

This time he swore he would get his legal help from outside the state of West Virginia. Charlie believes the entire state is controlled by the coal companies, including the lawyers who represent those companies. So he wanted a strong, independent law firm, "a law firm the company can't scare off or buy off."

Charlie, and many others in the Appalachian coal fields, had recently heard of the $11.5 million judgment Arnold & Porter had won for the disabled coal miners and miners' widows. So he called Harry Huge and asked him to come down. Harry went to the Valley, met with the members of the Buffalo Creek Citizens Committee, and then headed up to where the refuse-pile dam had been. At that point there was still a roadblock, and the state troopers would not let him through—until a committeeman said, "If the people's lawyer can't get up there to represent them, no more coal is coming out of this hollow." This is the way a "wildcat strike" starts in the hills of West Virginia. When there is something the people desperately need, they vote the only way that is effective—by striking and closing down the mines. The issue may not even involve coal mining. There have been numerous wildcat strikes in West Virginia over the content of schoolbooks, gas rationing, and almost anything else you can think of. The threat of a wildcat strike was enough to open the road. Harry got through, saw the dam site, and returned to Washington.

He had just returned when I called him. Together we decided to approach Arnold & Porter's Executive Committee to seek approval from the firm to represent the survivors of the Buffalo Creek disaster. The Executive Committee was called into session that very afternoon. Harry still did not know what it was we were being asked to do. We did not know how many survivors wanted us to represent them. We did not know how many of the survivors would be seeking compensation for the death of family or relatives, how many would be seeking recovery only for lost cars or houses, how many would be seeking recovery for injuries. We didn't even know whom to sue.

II

"An Act of God"

The dam which gave way was owned by the Buffalo Mining Company, a West Virginia corporation whose sole stockholder is the Pittston Company. After the disaster, Pittston, from its corporate offices on Park Avenue in New York City, proclaimed that "the break in the dam was caused by flooding—an Act of God." The people in West Virginia got the word, emblazoned across the front page of West Virginia's leading newspaper: "Flood Was 'Act of God,' Pittston Spokesman Says." Too much rain, too much water, run for the hills, the dam broke. As the country-music disc jockeys say, "The Good Lord willin', and the creek don't rise, we'll be seein' you next Saturday." But the Good Lord wasn't willing. You can't blame Pittston for that.

If this was a natural disaster, such as a hurricane or an earthquake, that would be an "act of God," in legal phraseology, and Pittston would not be responsible. Thus, Pittston's lawyers in New York probably felt it was good strategy to publicize the disaster as a natural one, an act of God. But telling the religious, God-loving, and God-fearing West Virginians that God had done this to them was a bad public relations blunder.

Robert Weedfall, West Virginia's state climatologist, responded immediately: " 'Act of God' is a legal term. There are other legal terms—terms like 'involuntary manslaughter because of stupidity' and 'criminal negligence.' "

The survivors in the Valley also reacted violently to Pittston's attempt to blame God for this apparently man-made disaster. One said, "You can blame the Almighty, all right—the almighty dollar."

Another added, "I didn't see God running any bulldozer. It's murder. The big shots want to call it an act of God. It's a lie. They've told a lie on God, and they shouldn't have done that. God didn't do this. He wouldn't do that."

Bill Davies, a federal official familiar with the history of coal-mining disasters in West Virginia, also couldn't agree with Pittston's New York lawyers: "God must be awful mean if he picks on West Virginia all the time."

Ben Tudor, a local Buffalo Mining Company official who wasn't privy to the strategy of Pittston's New York lawyers, tried to place the blame on West Virginia, rather than God, but that didn't work either. He implied that the state was at fault because it would not let Buffalo Mining Company drain the black waste water from behind the dams, for fear this black water would kill the trout stocked in Buffalo Creek. As Ben Tudor put it, the state "cares more for the fish than they did the people, and now both are gone." But a quick check of state records showed that the coal company never made any such request to drain its dam. So Ben Tudor clarified his statement. But his new statement, though more artfully worded for him, still implied that the state was at fault, rather than Pittston or Buffalo Mining Company.

There were many other confusing reports as to who was responsible for this disaster. *The New York Times* quoted

Robert Reineke, a Pittston lawyer sent to the scene of the disaster, as saying:

> The responsibility is Pittston's in the long range. The landowner would have no responsibility. I would say the refuse-pile is our responsibility. Before we investigate, we're not assuming liability. But we don't deny that it is potentially a great liability.

When Pittston's executives in New York read this, they had Mr. Reineke set the record straight, immediately, with this telegram:

> Today's "New York Times" (February 29, 1972) purports to quote statements I am said to have made to your reporter, Mr. George Vecsey. Among other inaccuracies of this report is a quote reading: "The responsibility is Pittston's in the long range." I never made any such statement to your reporter or to anyone else. I did say, as reported, that we were investigating. It would be absurd for me to express any opinion about anyone's responsibility before the investigation is completed.

Well, maybe it was "absurd" to speculate, two days after the disaster, "about anyone's responsibility" for the disaster. Still, the coal company's first written press release blamed "heavy rains" and "rising flood waters," a more sophisticated way of blaming nature without using those fighting words, "act of God":

> The heavy rains which caused rising flood waters to break through the dam at Lorado has brought the entire community to a standstill. Our company is doing everything possible to assist state and local authorities to care for the people involved in this unfortunate flood.

This press release also showed that Pittston was hard at work preparing another legal defense. Pittston's New York public relations firm issued the release in the name of the Buffalo Mining Company, Pittston's subsidiary West Virginia corporation, rather than in the name of the Pittston Company, the parent New York company. In this way, Pittston hoped any possible lawsuits would be filed against the Buffalo Mining Company rather than Pittston.

This would become one of the most critical legal issues in this case. If the West Virginia survivors sued the Buffalo Mining Company, a West Virginia corporation, their suit could be brought only in a West Virginia state court. Generally speaking, citizens of the same state may not sue each other in a federal court. On the other hand, the Constitution of the United States does permit citizens of one state to sue citizens of another state in a federal court. Thus, if the West Virginia survivors sued the Pittston Company, a New York corporation, their suit could be brought in a federal court.

Usually, it is the outsider who wants the protection of the federal court, since this constitutional guarantee of a federal forum for suits between citizens of different states was intended to protect the outsider. Here, it appeared that the outsider, the Pittston Company, preferred the state court, while the local citizens, the survivors, preferred the federal court. The reason was obvious—coal companies have more influence with the local West Virginia courts than they do with the less political federal courts, despite the fact that the coal companies are nominally the outsiders.

So this was Pittston's opening strategy—blame God and force the survivors to sue the Buffalo Mining Company. Our strategy would have to be—blame Pittston and ignore the Buffalo Mining Company. If a man worked for the Buffalo Mining Company, he was still a "Pittston employee" to us.

Meanwhile, others who might have been accused of responsibility for this tragedy were busy pointing the finger of blame elsewhere. The United Mine Workers, through its then president, Tony Boyle, rushed out a statement calling for federal legislation to prevent similar disasters:

> Our Union has warned on many occasions that these slag heaps are unsafe. We have urged legislation to clean up this kind of mess in the mine regions.

But there already was extensive federal legislation on the books. The House Subcommittee on Mines and Mining, which had drafted the Federal Coal Mine Health and Safety Act of 1969 after the disaster in Farmington, West Virginia, claimed the lives of seventy-eight coal miners buried by an underground explosion in a mine, immediately called for the appearance before it of the Secretary of the Interior, who had been charged by Congress with the responsibility for enforcing the act. Secretary of the Interior Rogers C. B. Morton declined the invitation. His brother, the former senator from Kentucky, Thruston B. Morton, happened to be a director of the Pittston Company. Secretary Morton avoided any embarrassment by sending Hollis Dole, the Assistant Secretary in charge of mineral resources for the Department of the Interior. Within three weeks after the disaster, Mr. Dole told the House subcommittee that the Department of the Interior felt it had no authority under the act with respect to the refuse pile or dam which failed at Buffalo Creek. If anything, he implied, Congress was at fault for not giving the Department of the Interior sufficient authority to make such facilities safe. He then urged Congress to pass some more laws.

Mr. Dole did admit there were specific federal standards which covered the construction of refuse piles and dams by coal companies. But he pointed out that the act's standards protected only the coal miner and not the public who lived around coal mines. Indeed, according to Mr. Dole, the act did not even protect coal miners when they were asleep in their homes or on the road to and from work—the coal miner was protected only while he was working on the coal-mine property. Mr. Dole added that "there were no employees on duty lost in this flood." So there would appear to be no federal responsibility under the Federal Coal Mine Health and Safety Act.

The congressmen who drafted this act naturally reacted strongly to Mr. Dole's attempt to excuse the federal government from responsibility for this disaster. They argued that the act was intended to make coal mining safe, to protect the coal miner and his family from coal-mining disasters.

Other congressmen, however, rushed to the defense of the Department of the Interior and the federal government, agreeing with Mr. Dole that the federal government had no authority or responsibility under its statutes to ensure the safety of refuse piles or dams. Instead these congressmen argued that the state of West Virginia, not the federal government, was probably at fault. As Congressman John Saylor put it:

> If anybody can read plain English, including all of Nader's Raiders, I would suggest that they go back and read that bill and the preamble of what Congress said and enacted and not try and expand it, taking care of a flood situation which in all probability was the direct result of negligence on behalf of state employees or persons in that area who were charged with the responsibility of erecting dams.

There was some basis for holding the state of West Virginia responsible for the Buffalo Creek disaster, since a specific West Virginia statute prohibits the construction of any dam or other obstruction over fifteen feet in height across any stream or watercourse without a prior determination by the state that it is safe. Everyone agreed that the state of West Virginia had never determined that the dam or obstruction which caused the Buffalo Creek disaster was safe. It could be argued then that the state was at fault for not enforcing its own laws.

But the governor of West Virginia was in no mood to

permit anyone to blame the state of West Virginia for this disaster. In fact, he himself had discovered the real villain in this piece—the news media. According to Governor Arch Moore, they were "irresponsible" in publicizing these attacks on the state of West Virginia—such as the Pittston accusation that the state cared more for the trout than for the people. As Moore put it:

> The only real sad part about it is that the State of West Virginia took a terrible beating which far overshadowed the beating which the individuals that lost their lives took, and I consider this an even greater tragedy than the accident itself.

Some even blamed the individuals who lost their lives for failing to heed alleged warnings that the dam might fail. For example, the Department of the Interior issued a press release headed, "Buffalo Creek Residents Had Prior Warning of Disaster, Interior Official Testifies." According to this official, Mr. Dole of the Department of the Interior:

> There had been at least four other occasions when the residents of Buffalo Hollow had been warned of an impending disaster which did not occur and, as a result, they were reluctant to move when the dam did fail. There are reports that warnings were given, that those who responded immediately were able to reach higher ground above the flood, but those who hesitated were lost.

III

"When in Doubt, Do the Right Thing"

It seemed clear the people needed legal help. Arnold & Porter's Executive Committee told Harry to return to the Buffalo Creek area to find out exactly what it was Arnold & Porter was being asked to do. Harry then met with the people in a mass public meeting at the Buffalo Creek Grade School, where he asked them to fill out forms which would tell us how many people wanted to sue for the wrongful death of a family member, how many wanted to sue for property damage, and so on. At the end of the meeting, Harry and his secretary attempted to complete the form with each person who requested Arnold & Porter's help. But so many people lined up, it was impossible to talk with each one individually. So some were allowed to take the form home to complete it and then return it to Charlie Cowan at his gas station the next day. Harry continued the next morning helping people who came in to the gas station to complete the forms. He then returned to Washington.

A few days later, Pittston, through the Buffalo Mining

Company, publicly announced that the Buffalo Mining Company would open offices in the Buffalo Creek Valley to receive and process claims arising out of the flood. They still refused to admit any responsibility for the disaster—"the investigation has not progressed to the point where it is possible to assess responsibility." The announcement did not even admit that any payment would be made to anyone filing claims. They would only "assess claims and evaluate losses." Nevertheless, the announcement that a claims office would be opened was read by some of us at Arnold & Porter, with a good deal of wishful thinking, as proof that our presence in the Valley had already accomplished a major goal on behalf of the people.

But Pittston probably would have opened claims offices even if Arnold & Porter hadn't come to the Valley. It has been traditional in mining disasters in West Virginia for the coal company to move rapidly to pay the widow or the injured miner. The money often comes from the insurance companies anyway, and insurance companies believe that the lowest settlements usually come when payments are made immediately after the disaster. At that point the victims are still somewhat dazed, and they usually have not contacted a lawyer or recognized the full extent of their injuries. They are particularly open to any kind of sympathy and want to believe, and thus do believe, they are being well taken care of by the coal companies and their insurance carriers.

Moreover, West Virginia state law makes the maximum amount payable in wrongful death cases so low that it doesn't pay for the heirs to refuse to settle. The West Virginia legislature limits recovery for wrongful death to $10,000, unless the family can prove greater financial dependence on the deceased. Even then, the maximum allowed by statute is only an additional $100,000. Since most of the deceased were housewives or children, Pittston argued that no one could be financially dependent on them. So Pittston claimed the state law maximum for each deceased woman or child was $10,000, rather than $110,000.

The people in the coal fields realized that Pittston's willingness to open a few claims offices was not a great victory. Pamphlets and newsletters from various organizations warned the victims of this disaster not to be taken in by the opening of the claims offices and to be wary in any negotiations with the Pittston lawyer who was to head this office. For example, *Coal Patrol,* published by Appalachia Information, said:

> In the midst of confusion and rumor, Pittston's lawyers have made sure that Buffalo Creek residents hear about the state's courts—and how the state law protects companies against individual judgments larger than $10,000. The idea seems to be that the people in desperate need of money will despair at the thought of long courtroom battles ending up—after years of appeal—with piddling awards. They'll be inclined instead to take whatever the company offers. Of course, the company hasn't offered *anything* yet—but Pittston has set up three "claims offices" along the Buffalo Creek and is encouraging people to drop in and let the company know how much they've lost, what they might want to settle for, and other details like that. Reportedly some people have also been asked to sign agreements releasing Pittston from any liability above and beyond the initial settlement. The company says it's not doing anything like that—but that's been the normal practice after mine disasters (insurance companies rushed in to get similar waivers from the widows of men killed at Farmington in 1968 and Hurricane Creek in 1970), and there's ample reason to be skeptical about Pittston's real motives behind the "claims office" front.

Pittston's motives became even more apparent when it publicly insisted that the survivors should not bother to hire a lawyer before filing their claims, because Pittston would not pay any more to survivors represented by a lawyer than it would to survivors who came in on their own.

Back in Washington, we were reviewing the forms the people had filled out requesting our assistance. Two hundred or so completed forms indicated that most of the people seeking our help had lost their homes or property, but no members of their family. In general, those who had lost family either had gone to lawyers in Logan County or were still too upset to even consider talking to lawyers about their claims. The Executive Committee at Arnold & Porter met on April 6, 1972, to hear our plea that the firm represent the two hundred people who already had requested our assistance and any other such people who might join them.

This was probably the worst possible time to be asking Arnold & Porter to take on the representation of such a large number of people in what promised to be such a big lawsuit. Arnold & Porter was as busy as it had ever been, at least during the time I had been with the firm. The firm's assignment committee reported to the Executive Committee that there were no lawyers available to work full time on any new matters and that it would be impossible, with the firm's present workload, to staff a massive new lawsuit. Some new lawyers had been hired from that year's law school graduating class, but they would not be joining the firm until the end of that summer. Harry could not break free from his other work to spend full time on the case. But I could. I said I would devote the remaining half-year of my *pro bono* partnership exclusively to this case and assured the Executive Committee I would not need any large number of lawyers to assist me until we were ready to file the lawsuit. But the manpower problem was still a serious one.

In addition, since the survivors could not afford to pay a legal fee to Arnold & Porter, we would have to represent them on a contingent-fee basis. If they won their lawsuit, our firm would earn a percentage of their award. If they lost, we would get nothing. Realistically, we could assume we would collect some award for the people, but the size of the award might not provide a contingent fee large enough to pay for the hours of legal time we would have to put into preparing and trying their complex lawsuit. Moreover, Arnold & Porter had not taken a contingent-fee case for many years. So there were a number of good and valid reasons why the Executive Committee might decide not to represent these people in West Virginia.

But Paul Porter, one of the founders of the law firm, has a favorite saying—"When in doubt, do the right thing." Because of the public interest aspects of the case, Arnold & Porter agreed to represent the survivors. Harry and I made plans to go to the Valley.

IV

"Mountain Mama"

I was a little apprehensive before my first trip to West Virginia. To me, coal mining and West Virginia meant death and violence. Jock Yablonski, Tony Boyle's opponent for president of the United Mine Workers Union, had just been murdered, along with his wife and daughter. Logan County, where Buffalo Creek is located, itself has a long and bloody history of violence, from the Hatfields and McCoys (the Hatfields ended up in Logan County, West Virginia, the McCoys across the border in Kentucky) to the bitter Logan County war in 1921 between armed union organizers and the Logan County sheriff's army paid for by the county's coal operators.

I'd had these same kinds of fears—of violence, of the unknown—during the 1960s before I made my first trip into the backroads of Mississippi, Alabama, and Louisiana to help blacks who'd been denied the right to vote. That was back when Kennedy was President, Martin Luther King, Jr., was jailed in Birmingham, James Meredith was trying to stay alive at Ole Miss, and Medgar Evers was killed on his front porch in Jackson, Mississippi. To an outsider, those seemed like scary days to be going to the South. But once I'd gotten used to flying down and returning alive, the violent South of the newspapers seemed less frightening to me. I hoped I'd also get over these fears of West Virginia, once I'd been down and back a few times.

I was also a little anxious about meeting the survivors. I knew from my civil rights days that people beaten down by the system don't open up easily to strangers. It had taken me many, many trips to Mississippi to talk with the black people, in the fields, on their tractors, behind their mules, in their homes, at their churches, or wherever I could corner them, before they would start to trust me and tell me even part of their real story. I never have forgotten George Cotton, in Clarke County, Mississippi. He taught me that it takes persistence and patience to gain another man's trust, when that other man is afraid to speak out.

I'd subpoenaed George Cotton to testify in federal court about the times the Clarke County registrar wouldn't let him register to vote. George Cotton didn't make it to that hearing. He ended up in the hospital. I learned much later that one of the county powers had been by to tell him his loan might be called in if he went to the hearing. So George decided to fall off his tractor and spend the day of the hearing in the hospital.

I was upset when he missed the first hearing but decided to try again for the next one. I handed him another subpoena. He said he'd try to come. I looked him in the eye as we sat on his porch, grinned a bit, and told him I really wanted him to come. He put his hands in the pockets of his overalls, looked my way, and said, "I'll try and be there." I finally told him, "Mr. Cotton, let me read this subpoena to you. It says 'You are hereby commanded to appear'—it doesn't say 'please.' But I'm asking you, please." He laughed, and came. Later, after many more visits, and the final trial had ended, he told me, "I didn't feel I could trust you at first, but about the second time after you gave me that subpoena I knew I could rely on you. It made me feel better each time you came by."

I felt Harry had earned this kind of trust from the West Virginians he'd worked with on his Welfare Fund case during the past three years. I hoped it wasn't going to take too long for me to earn that same kind of trust from the West Virginians we were now going to see.

The day before Harry and I were to fly to West Virginia, we learned that one of Logan County's few plaintiff's lawyers, Amos Wilson, was in a fury over Arnold & Porter's decision to represent survivors of the Buffalo Creek disaster. Mr. Wilson often represents Logan County's coal miners in workmen's compensation cases and black-lung cases. Since he was the leading plaintiff's lawyer in the county, he felt he, not some out-of-staters, should represent these plaintiffs. He was even threatening to bring a lawsuit to enjoin Arnold & Porter from representing any of the survivors of the disaster.

We really had not thought through what the consequences would be if Amos Wilson did go to court to get an injunction prohibiting us from representing any of the survivors of the Buffalo Creek disaster. However, we were certain that Arnold & Porter had done nothing unethical, since the Supreme Court had recently held that organizations such as the Buffalo Creek Citizens Committee have the right to recommend particular lawyers to their members. I wasn't so sure though, that the law as interpreted by the Supreme Court had filtered its way down to Buffalo Creek in Logan County.

Buffalo Creek is a long way from Washington, D.C. It is at least an eight-hour drive through and over the Blue Ridge Mountains and then along winding Appalachian roads in West Virginia. West Virginia has always been difficult to reach by road because of its mountainous terrain.

And it's not much better by air. Charleston, the state capital, is only an hour's flight from Washington, but you often don't know when you get on the plane in Washington whether the plane will be able to land in Charleston. The fog rolls in rapidly and often causes pilots to bypass Charleston for a safer airport, sometimes as far away as Cleveland or even Chicago. Moreover, the Charleston airport was built by leveling off the top of a mountain, and a small mountain at that, so the runway is too short, and the landing is always unnerving.

But we made the flight without any problems. In Charleston we rented a car for the two-hour drive to Buffalo Creek. Within an hour from Charleston, we began to enter coal country. As we drove through the narrow hollows and valleys, towering mountains on each side, it was almost impossible to see the horizon. Maybe this is the reason people in West Virginia have accepted their fate with the coal companies for so many years. It is easy to lose hope when you can't even see beyond your own valley.

On the other hand, I remember once picking up a hitchhiker on one of these narrow valley roads. He told me his first sight of the ocean, with the horizon far off from view, made him queasy. He said he was happy to get back to his West Virginia valley and the comfort of its mountains. In the Buffalo Creek disaster, these mountain walls meant safety and life for many of the Valley residents who ran up them before the waters roared through.

So the enveloping mountains may shut out the horizon and the outside world, but sometimes that's a kind of peace to be desired. John Denver sings about this comforting Mountain Mama in the song about West Virginia:

West Virginia
Mountain Mama
Take me home
Country roads.

We drove on down these country roads, through and over the mountains, behind slow-moving coal trucks, and finally reached Buffalo Creek. Six weeks after the disaster it still looked like a war zone. The National Guard was everywhere, bulldozing destroyed homes into big piles for burning, directing traffic over temporary wooden bridges, searching the rubble for bodies. The U.S. Army Corps of Engineers had its big machines down in the creek, clearing out the debris and widening the channel. Houses here and there were marked with a large X, meaning they soon would be leveled and burned. The railroad crews were hard at work putting in new rail lines. Black water marks showed everywhere, on stores, churches, houses—even on the side of the hill. There was an eerie nothingness on each side of the road, what was left of it. Where towns had been, there now were only railroad signs, announcing that here once stood Latrobe or Crites or Lundale. As we drove to Charlie Cowan's gas station, the smoke from the burning homes filled the narrow valley.

Charlie was dressed in his work clothes and seemed very stand-offish. I was eager to make a good impression, but we hardly conversed. I was surprised to see he had lost most of his teeth, but as I got to know West Virginia more I saw this was a common phenomenon of the mountains, where dental care is either too expensive or nonexistent. After a while Harry and I, and a few miners, drove up the Valley to visit the site where Pittston's dam had been located. At Saunders, where Middle Fork emptied into Buffalo Creek, we got out.

We walked way up Middle Fork Hollow, at first on the side of the hollow over a massive black refuse pile, and then through mud and sludge where the miners said three dams had blocked the stream which ran through Middle Fork. We were walking on a site many times the size of a football field. There were no people around, no homes, no dams or impounded water, just a little stream running down Middle

Fork through black oozing sludge. If you stepped in the wrong place, the sludge pulled you in like quicksand. I could see, here and there, pockets of fire in the remains of the refuse pile. And some parts of the three dams still jutted out from the sides of the hollow.

Months later, with the help of aerial photos and expert studies, I slowly began to understand what had been there—a large burning refuse pile on the side of Middle Fork Hollow and three dams of coal-waste refuse built across the hollow. The refuse pile was started back in 1947 when the coal-cleaning plant first opened. A coal-cleaning plant, or "tipple" as the miners call it, is an integral part of any coal-mining operation. The coal comes from the mines by conveyor belt and goes directly into the tipple, where it is washed and graded. The clean coal drops into railroad cars underneath the tipple for delivery to customers. The tipple gives off an enormous amount of waste products—solid refuse, 800 to 1,000 tons a day of slate and rock and coal waste; and especially liquid refuse, 400,000 to 500,000 gallons of black water each day containing about 500 tons of solids. For years the liquid refuse was discharged directly into Buffalo Creek while the solid refuse was trucked to the mouth of Middle Fork and dumped on the side of the hollow.

This dump, or refuse pile, grew almost continuously. By the time of the disaster it was a smoldering mass of 3 million cubic yards of refuse, over 2,000 feet long and over 400 feet wide. It burned constantly because of the waste coal compressed within it.

Sometime in the early 1950s West Virginia told the coal companies to stop discharging the black waste water from their tipples directly into the state's streams. So the Buffalo Mining Company tried pumping this sludge into old mined-out areas underground or into dugout ponds above ground. Then in early 1960 the company decided to build a dam across Middle Fork to provide a settling area for the black water. This dam, known as Dam 1, was built by having the trucks dump some of the solid refuse across Middle Fork Hollow, instead of just along the side. Dam 1 was really only a small dike built at the upper end of the existing refuse pile. It was 6 to 8 feet high, 15 to 20 feet in length up and down the hollow, and 100 feet wide across the hollow. A pipe from the tipple then discharged the black water at a point far upstream from Dam 1 so this water could filter through the refuse of Dam 1 before entering Buffalo Creek at Saunders.

This marriage of convenience, using the solid refuse to filter the liquid refuse, worked fairly well until the solids in the liquid refuse silted up behind Dam 1. So, in 1967, Dam 2 was begun, upstream from Dam 1. Again, the trucks merely dumped the solid refuse across Middle Fork Hollow until the refuse reached the other side. Dam 2, 88,000 cubic yards of refuse, was 20 feet high, 25 feet long, and 450 feet wide. Dam 2 soon began to silt up too.

So Dam 3 was started upstream of Dams 1 and 2 in 1968. By the end of 1970 Dam 3 reached the other side of the hollow. By the time of the disaster its 534,000 cubic yards of refuse reached almost 60 feet high, extended 400 to 500 feet up the hollow, and spanned 450 to 600 feet across the hollow. The water in this dam towered almost 250 feet above the Saunders at the mouth of Middle Fork below. It was the failure of Dam 3 which caused the Buffalo Creek disaster on February 26, 1972.

After our visit to Middle Fork, Harry and I went to the Buffalo Creek Grade School to meet with the survivors of this disaster. The gymnasium floor was crowded with metal folding chairs, and every chair was taken. People were also standing around the walls of the gymnasium and were jammed in at the doors. This small gymnasium was completely filled with over one hundred families of survivors of the Buffalo Creek disaster. Harry and I decided we wanted no outsiders or news reporters present at this meeting when we began our talk to the people. Some of the more burly miners escorted the outsiders out of the meeting. Harry then told the people that Arnold & Porter had agreed to represent them and that I would be the lawyer in charge of their case. He added a little bit about my background as a litigator for the federal government in civil rights cases and then turned the meeting over to me.

I was a little nervous as I looked out at their faces, everyone desperately hoping we would find the way to put their lives back together, and soon. I had to tell them the truth. I had to make it clear that we could not offer any hope of immediate success in any lawsuit against Pittston. If they sued Pittston, they had to know they'd be in for a very long fight. If they could not afford to wait for years to get their money, or if they were in such desperate straits that they had to have some compensation now, it might be in their best interest to go on down to the Pittston claims office to see what they could get. They could always come back and join our lawsuit if they felt Pittston was not going to pay them enough.

I probably painted an overly bleak and pessimistic picture. But I remembered how uneasy I was in my days in the Civil Rights Division when black families we visited looked upon us as some kind of messiah. I recognized then how unusual it was for any family, and especially a poor, helpless black family in the South, to have a lawyer from the federal government in Washington come to their home and ask if the federal government could help enforce their constitutional rights for them. No government lawyer has ever knocked on my door to see if the federal government could do anything for me. Indeed, if the government ever did knock on my door, I'd probably expect harm and harassment instead of help. In the South, though, our many visits finally convinced these people we really did mean it when we said we were coming to help them. But once that feeling got across, and they let their defenses down, I became concerned that we might have held out greater promise than we could fulfill.

Here in West Virginia, there were no defenses. These people opened up at once to our offer to help. We did not come as agents of the federal government, so there was no fear in that sense. We also came highly recommended by people they themselves trusted. As a result, I immediately tried to downplay what it was that we might accomplish for them, not only to keep from raising their expectations too high, too soon, but also because I did have great concern as to what it was we really would be able to do. We were already under attack from Amos Wilson for coming in to represent them. And my experience with judges in the South had soured me on the chances of using the courts to accomplish anything worthwhile. The courts usually are a great place to delay matters, but no place to go if you need help in a hurry.

We also spent some time, during the meeting, discussing legal fees and costs. Before going to the Valley, I had drafted a form retainer letter to set out in writing for each person exactly what the legal fee would be under the contingent fee arrangement. In cases such as this, lawyers usually charge a contingent fee of 33-⅓ percent or even more. We decided to make the contingent fee around 25 percent, plus expenses, realizing that if we made the percentage too low we would be giving further ammunition to Amos Wilson, who already was objecting to Arnold & Porter's presence. If the fee was too low, he might argue we were violating the lawyers' Code of Professional Responsibility by taking a case for free when the client could afford to pay another lawyer. The draft retainer letter also set forth the way in which the various costs of the suit would be allocated and paid. We would pay all those expenses, but the clients would have to pay us back. Of course, if we lost the suit, it was hard to imagine us suing the plaintiffs to collect for the expenses they owed us.

I spent the next few days calling on Logan County's lawyers to try to defuse any lawsuit to prohibit us from representing the survivors, and to try to learn a little about the rules of coal-field warfare.

Ed Eiland is one of the leading coal-company lawyers in Logan County. He had just lost the election for judge of the Logan County Circuit Court, so I assumed he also understood the politics of the county. I told Mr. Eiland that Arnold & Porter probably would represent a number of the survivors of the Buffalo Creek disaster. He saw no objection to our representing these people, since he felt no single practitioner in the county could possibly handle a case of the magnitude we were discussing. He also told me that there were few Logan County lawyers available to handle personal-injury actions. He added that he had been approached to represent one of the potential defendants in the case, the land company which owned the land on which the dam was located, but he had a possible conflict of interest because one of his coal-company clients might itself be a plaintiff suing for the property damage which it suffered from the disaster. Mr. Eiland did tell me that Amos Wilson was upset that Arnold & Porter would be representing a number of the survivors, but he added that he thought Mr. Wilson could not take on a case this big.

Mr. Eiland had suggested that I talk with the chairman of the Logan County Bar Association and with another Logan County lawyer who happened to be one of the ten members of the West Virginia State Bar Association Ethics Committee. So I met them. The chairman of the Logan County Bar Association told me he did very little plaintiffs' work of the sort raised by the Buffalo Creek disaster and that he had no personal concern with Arnold & Porter's involvement in the matter.

The Logan County lawyer who is a member of the West Virginia State Bar Ethics Committee also told me he had no concern with Arnold & Porter or any other out-of-state law firm representing claimants in the Buffalo Creek matter. But he added that there was a plaintiff's lawyer in the county who had complained to him and to the chairman of the West Virginia State Bar Ethics Committee. He said Arnold & Porter had no reason to be concerned, unless Arnold & Porter had been soliciting business. He then told me how upset a number of West Virginia lawyers had been when an out-of-state lawyer flew into Huntington, West Virginia, immediately after an airplane crash occurred there, to begin soliciting business from those survivors.

Next, I called on Logan County's prosecuting attorneys. In West Virginia the prosecuting attorneys are permitted to have a separate private law practice, apparently on the assumption that the counties cannot afford to pay lawyers a sufficient amount of money for them to work full time for them. Logan County's three prosecuting attorneys had already been privately retained by some survivors. This later proved to be an obstacle to anyone who might have wanted Logan County to bring criminal prosecutions against the Pittston Company. Since Logan County's prosecuting attorneys were already representing people in civil lawsuits for damages against Pittston, legal ethics prohibited them from instituting or preparing any criminal proceedings against Pittston. It would be quite some time before Logan County officials gave any serious consideration to determining whether anyone was criminally at fault for the death of over 125 people.

Doug Whitten, one of the assistant prosecuting attorneys, was hard at work, in his capacity as a private lawyer, drafting a civil complaint against Pittston. He said he would sue Buffalo Mining Company in the local Logan County state court, rather than in the federal court in Charleston, if he couldn't get a fair settlement for his clients. He said he preferred the local juries in Logan County rather than the juries in federal court over at Charleston. He also said he was happy Arnold & Porter would be coming in to represent a number of the survivors, since this would help him with the legal research for his lawsuits. Oval Damron, the county's chief prosecuting attorney, was not so pleased. But he didn't threaten any action against us for agreeing to represent some of the survivors, so I felt somewhat relieved.

It was time to go and meet Amos Wilson. He had the most elaborate office of any of the Logan County lawyers I visited. He had a whole floor upstairs over the dry goods store and across the street from the County Courthouse.

We quickly got past the pleasantries. I told him Arnold & Porter had prepared a draft retainer letter for those who had asked us to represent them and that we would probably represent a substantial number of claimants in the Buffalo Creek disaster. He said he had proof in his safe that Arnold & Porter had already obtained signed retainers from some people. I said that wasn't true, yet.

He said he felt he was entitled to at least one-third of any of the business from this disaster in his own home county, since he was the county's leading plaintiff's lawyer. He was quite upset that Arnold & Porter might be retained by a large number of survivors. He said they might have come to him instead if we hadn't come on the scene.

He complained about the Appalachian Research and Defense Fund (APPALRED), the public interest law firm which had provided free legal advice to the Buffalo Creek Citizens Committee and had helped in the creation of that committee. He charged that after the disaster, APPALRED's lawyers were permitted to get by the roadblock and up into the Valley, while he had to wait. He told me he felt APPALRED had gotten the people to call in Arnold & Porter.

He also said he was informed that Arnold & Porter would be giving APPALRED a portion of its contingent fee in violation of the regulations which prohibit public interest law firms from earning money as lawyers. I tried to assure him that Arnold & Porter would not be paying a referral fee or any other fee to APPALRED or any of its lawyers. I don't think he believed me.

Mr. Wilson did admit that, by himself, he could not represent all the people in Buffalo Creek. But he said he had associated himself with two law firms in Charleston with a total of fourteen lawyers. He felt his group could represent everybody, and there was no need for outside lawyers. That ended our conversation. I was sure he would continue to oppose our representation of the survivors.

I went back to Washington to prepare individual retainer letters for over 200 adults, who, with their children, represented over 450 survivors of the disaster. We were beginning to realize the enormity of representing so many separate plaintiffs. Just getting the names right on each typed retainer letter took a week. I returned to Charleston and drove to the Valley. Handing the formal retainer letter to each plaintiff depressed me. Their needs were so great, and we had such a long, hard battle ahead of us. And all they'd gotten from me so far was a lengthy, complex legal paper to sign.

After completing the personal delivery of the letters, I drove the wearying trip back to Charleston. I slumped into a chair at the Charleston airport and picked up a newspaper from the next chair. The headline read, "Buffalo Creek Lawyer Acts Due Look." The Kanawha County Bar Association (the bar association for the county in which Charleston is located) had just voted to investigate "the alleged solicitation of legal business" by Arnold & Porter. According to the paper:

> Behind the Kanawha Association's decision to investigate is not only the loss of lucrative business to an out-of-state firm, but also the conviction that Arnold & Porter can't do as good a job for its clients as lawyers familiar with West Virginia law, said a half a dozen Charlestown lawyers interviewed Friday.

I may have protected us in Logan County by my calls on Logan County's lawyers. But now it appeared the Charleston bar was after us. I wondered if Pittston, with its influence in Charleston, might be behind this investigation. If so, this was a bad sign.

V

"Take Care of My Baby"

Now back and forth. For four months. Drive to National Airport in Washington. Get on the Piedmont flight. Get off at Charleston. Rent a car. Better rent a compact this time. I can't keep those big cars on the road. Tires keep slipping over the narrow shoulders. Drive to the Valley, over Kelly Mountain, past Island Creek's coal tipple, past some burning refuse piles, under the coal tramway, past Amherst Coal Company's tipple, and on to Charlie's station. Park in the mud and go to our "office."

I had decided to use a storage room at Charlie's gas station as our office. I turned down an offer from the Methodists to use their Wesley House building as our office because Charlie told me some of the non-Methodists might feel a bit uneasy going there. I never did know if that was true, or if Charlie just wanted to keep better tabs on us. But his station was more conveniently located for the people, so it became our stopping point.

Charlie's gas station was a small, white cinder-block

building with black water marks from the disaster still clearly in evidence near the roof. The waters had torn off the station's back door and ruined most of the food and candy stocked on the walls. But Charlie's shelves were full once again with cigarettes, all kinds of candy, bread, pies, cupcakes, potato chips, toys, and car gadgets. He'd even salvaged some old fishing gear for the trout which managed to live in the polluted waters of Buffalo Creek for a few days each year after the state fisheries stocked the creek. His coolers were working again, and they too were full to the top—soda pop, juices, milk, and ice cream. His little station office was a veritable child's delight, the hangout for kids of all ages, and an oasis for the miners who constantly stopped to gas up their huge coal trucks and grab some cigarettes, an armful of Cokes, and some candy for the rest of the men back at work.

Behind this office was the much smaller storage room, packed to the roof with cartons of Cokes and boxes of candy. It also housed a massive black safe, over six feet tall, and a bulky gray compressor which compressed the air for the grease rack and air hose. Charlie had added a filing cabinet, a small table, and an old wooden office chair whose sloping seat always threatened to deposit me on the floor. There was one bare light bulb hanging from the ceiling.

Often, while I sat in that crowded room talking with these suffering people, the compressor would suddenly come on. In that small room, its noises and hissing sounds were unsettling. To these people, still unnerved by the horror they had lived through, it was terrifying.

The people would sit on a little blue canvas beach stool or on Coke cartons, and I'd begin to get some of the details of their stories. "Please tell me everything you lost. Well, tell me as much as you can remember. I know it's hard. I wish there were some other way, but we've got to do the best we can. Take your time. Maybe you can make a list and send it to us. Where were you when the dam broke?

Then what happened? I'm sorry. We can go over that some other time if it's too difficult now. Well, how much did you pay for your house? Who did you buy it from?" On and on, all day.

This was my first real opportunity to meet and talk with the people individually. Until then I had addressed them only as a group in the gymnasium as I stood behind a long table. Now we were sitting knee-to-knee, and I was beginning to see the serious personal consequences of this disaster. They were crushed. Their whole demeanor demonstrated how overwhelming this disaster had been. It was hard for them to sit up straight or to talk for long periods of time without drifting off in their thoughts or without averting their eyes from my glance. Tears came quickly and often.

Roland Staten and I sat huddled together in the office. I leaned over, straining to hear his words. He had lived in House No. 20 in Lundale, West Virginia, a small coal-camp community of a hundred or so coal miners' families. Years ago, when Lundale and the other camps on Buffalo Creek were company-owned towns, the coal companies owned all the coal-camp houses. But when the union stopped the coal companies from raising the rents any further, the coal companies decided to sell the wooden frame, "four walls and a roof" houses to the miners. The miners took great pride in turning them into real homes, helping each other, or even paying someone to do the work once they saved enough money. An indoor bathroom, maybe new electrical wiring, electrical baseboard heating, new floors, a new roof, new siding to keep out the cold, maybe a new porch or even a new room. Roland and his wife, Gladys, spent seven years remodeling House No. 20—adding a cesspool, paneling, insulation, siding, a new roof and furnace, and even a garage. This was no coal-camp house anymore.

Buffalo Creek's coal camps weren't really camps anymore, either. They still went by the names given them by the coal companies, from the head of the hollow at Middle Fork to the mouth of the hollow at Man—Saunders, Pardee, Lorado, Craneco, Lundale, Stowe, Crites, Latrobe, Robinette, Amherstdale, Becco (Riley), Fanco, Braeholm, Accoville, Crown, and Kistler. But the sameness and desperation of the old company-owned coal camps were gone. Before the disaster, these camps bristled with life, with the vitality and excitement of a community of working people owning and fixing up their own homes and lives.

Roland was sleeping late that cold morning, February 26, 1972. It was Saturday, and the mines were not working. Gladys was asleep in bed beside him. Kevin, their baby, not yet two years old, was sleeping in the next room. The house was small, only five rooms and a bath, but it was big enough for the three of them, and the fourth, who was expected in just a few months. Then—"people screaming, and the roar of the water" shattered the Buffalo Creek Valley. Roland bolted out of bed for the window. There was a wall of black water headed for his house. He still couldn't wipe away that memory, any of it, months later.

"I ran back to the bedroom and put on a pair of pants and went to the other room and grabbed my son and I picked him up and by that time the house behind mine had bumped up against mine and my house started to move. My wife was in her nightgown and my son was in a nightshirt and a Pamper.

"We started to the left side of the house because there was no water at that time, you know, coming, but by the time we got to the living room, from one room to the other, I was standing in water up to my neck with my son, and I told my wife, I said, 'We've got to get out of here.'

"As the house behind mine hit mine the windows shattered in mine, and I climbed out the window to the left of the house. As I was doing that I cut my foot. It was an aluminum window and as I was getting out—my garage was on that side, and as I was standing in the window my garage left. It was gone. I was just sort of trying to hold on to the top of the window there, and as I was climbing out the water just sort of picked me up and I was on the roof before I even knew it with that boy of mine.

"My wife, she was hanging on the edge of the roof, and she—as I tried to help her up, she was kind of heavy—she was about five and a half months pregnant, and she was a big-boned-like woman, and well, she wasn't on the real heavy side, but she was heavy anyway, you know.

"And I picked her up or tried to pick her up with my left hand and holding my son in my right hand. And he was screaming and he knew something was wrong. He was screaming and carrying on, and as I tried to pick her up, why, I just lost my grip, you know, just the roof, the gable of the house, the way it was made, and her pulling on me too and I went back in the water with him in sort of a lurch, you know, and my wife says, 'What are we going to do, what are we going to do?' And I said, 'Just hold on to anything you can find, anything.' And by that time the water was so deep and so much force, why, I was twenty or thirty feet from her.

"When I looked back and saw her she said, 'Take care of my baby.' And by that time I was gone. That's all I heard. That's the last time I saw her.

"And I carried him down through there and my vision—I couldn't see because of the black muck and stuff, it was just blurred. Everything was just blurred, but I still had that boy of mine and we were going under and under and I was trying to hold him and keep him up and keep myself up at the same time.

"We were just thrown from side to side and I was just grabbing onto a center of somebody's tire that came off a car or something—it was inflated and everything, and I grabbed the center of it and I held onto it and just where it went that's where I went. And I was thrown from side to side and crushed—my insides was crushed so hard that it just seemed like my eyeballs was trying to pop out, and my breath, I just couldn't get my breath at all.

"Somewhere along there I lost that boy of mine. I don't know where.

"By that time he had stopped screaming and he had drank so much water and everything—I don't know what happened to him."

Roland continued to be crushed by the "wood and debris and automobiles and everything" as the black waters rushed him down the Valley. He was holding onto a tire, but he soon realized that "I was going so fast and bobbing along that I knew I had to turn loose of that tire." He did, just in time to miss a water pipeline that was across the creek—"how I got by the pipe, I don't know." He then grabbed hold of a ground strip from a refrigerator frame, but "it gave a slip and I don't know where it went. Then I was just in the wood, just mixed up in it and still going from side to side and not knowing where I was. I couldn't see."

Then, at a point where the Valley widened, the water spread out and "sort of lowered down." Roland fought his way over to a car body which was stuck in an old refuse pile there. "As I was going by I grabbed a spindle, where the wheel fits on, and I held on. I just couldn't breathe at all. I was just worn out, holding on with everything I had." Finally he managed to get over the car and climb up out of the water on to the "red dog" refuse pile.

"As I was laying there I felt my arms and body and I couldn't feel a thing, I was so cold, and while I was fighting my way to that car body I noticed this stick that was in me. I knew something was sticking in my side there. There was so much wood and everything there that it just kept pressing and it just kept going and I couldn't get it out, you know, and I just jerked it out to the side and when I did that it left quite a hole there.

"I laid on that red-dog pile until somebody—I don't know how long I laid there, I don't have any idea, but some people came and got me and wrapped me up in a blanket and took me to the church up on the hill, the Church of God. And I stayed there and they kept me wrapped up in blankets, and

I know a lot of people were there because they were talking. I couldn't understand what anybody was saying. They poured coffee down me and asked me anything I wanted, anything they could do, and I just didn't have an answer. It just seemed—it just didn't seem real."

It took less than two hours for the full force of the rolling tidal wave of black water to tear through the sixteen communities and seventeen miles of the Buffalo Creek Valley. Many people drowned immediately as the waters demolished their homes. There was absolutely no trace left of Saunders at the mouth of Middle Fork. Its twenty homes and the Freewill Baptist Church were gone.

Others survived the first onslaught of the waters only to die when their homes or the other pieces of debris they were clinging to broke apart as the waters dashed them into railroad trestles and highway bridges which crossed the valley floor. Jesse Albright is a tough miner, and he looks tough, too, with his big wad of chewing tobacco always bulging out his cheek. He made it to the safety of the hills, but he remembers the women who didn't.

"I turned and run up Proctor Hollow. I went up past the old electric substation. I turned around and I seen the first house hit the bridge, then I seen the second house hit, then the third and fourth. They just kept piling up. It seemed like after the houses all hit the bridge, a mobile home came floating behind them.

"There were three women standing in it, in a picture window, a big window. They were standing in the window and I saw their mouths moving. I gathered they were hollering.

"Now where the houses done jammed up against the bridge the mobile home hit those houses and I guess the pressure and the impact was rolling under and that mobile home with the three women in it just vanished underneath all those houses and I never did see no more of the mobile home."

Although many reached high ground before the rampaging waters destroyed their homes, the sights they saw still haunted them. They had to stand by on the hills and watch helplessly as friends and neighbors, still alive, were swept downstream by the torrent, clinging to whatever wreckage they could, screaming for help. James Morgan mined coal for twenty-six years before he hurt his back and had to quit. "I seen two little kids going down on a mattress just before the water hit us. They were bloody all over and hollering for help, 'Hey, mister, help me. Help me, mister.' And there wasn't nothing I could do."

Leroy Lambert had only one good leg. He'd lost his other in a mining accident. And now he'd lost his crutches, but he still made it up the slate dump behind his home before the water came through. "When we was going up the slate dump we heared a scream and looked back and it was Jason Bailey's little boy screaming for help, but there wasn't no way we could get to him because he was thirty or thirty-five or forty feet out in the water."

Rex Howard started in the mines in 1929, but black lung and a broken back finally forced him to retire in 1971. He couldn't help the children—"There were so many of them. I heard the kids screaming when they went by the house."

The survivors saw countless dead bodies floating down the creek, and their homes and all their personal belongings being carried away by the waters. As Ora Mae Hagood, a teacher, remembers it, "We were running, scared, trying to find a place to go to where we thought we would be safe, and then the terror, the horror after we came back off the hill.

"During the initial shock, the time of the water coming down, we just stood there so helpless. We just stood and stared, nobody talking, just like we were in a trance, just numb. A house would go by, a car would go by, and you would wait for yours to fall down and go by. You'd see your neighbor's house go by, everything that they had worked for for so long.

"Then you began to wonder was anybody in the house?

Did everybody get out? All kinds of thoughts. We had no idea how much force that water had.

"And we stood there just so helpless, couldn't do anything.

"I remember one remark. They said, 'Well, the post office has washed all the way down to the postmistress's house.' And they said to the postmistress, 'You won't have to go to the post office any more because the post office is right in your door.' Just anything to try to keep ourselves from going to pieces.

"At first we couldn't cry. We couldn't cry. We were just appalled at the horror, the debris, the things that were coming down; two-story houses, just like dominoes, one after another, just like a chain reaction set off."

Then, as the waters receded and the bodies of the dead became apparent, they saw even more gruesome scenes of death. Again, Ora Mae Hagood: "The first thing when we came off the hill there was a dead body beside our house, up above the road by our house, and we were called to see if we could identify the body. I think they thought perhaps we could identify the body because my husband taught high school and I taught in grade school and one of us possibly could. But we couldn't identify the body because the body was so black. We couldn't—we just couldn't—it was so horrifying to look at it.

"They thought perhaps that she was a black person. Come to find out later on all the people looked like that, greasy and black, and the look on their face was just horror. If you've ever seen anybody die a violent death, it's not like going to a funeral home and seeing people all dressed and prepared. Instead, you see the fear in somebody's open eyes, mouths awry, it's just a horrible thing."

In many cases, the survivors at first could not even identify members of their own family because the black coal waste made them unrecognizable.

For the survivors, their ordeal had just begun, and the

days and weeks that followed exacerbated their already precarious and fragile state. Many of those who reached high ground gathered together in undamaged houses. The people remember the exact number who crowded together with them in various houses on the hill after the disaster. They counted each other over and over to reassure themselves they were alive. There were thirty-seven in Herman Stiltner's house. Seventeen in the four-room house where Ethel Sparks went for refuge. Nineteen in Charley Walls's house up one of the side hollows.

There were forty-two where Doris Mullins was, not counting one old woman, pulled from the water alive, who died in agony before their very eyes. As Mrs. Mullins recalls it, "She was choking from the black water and it appeared that every bone in her body was broken. Her body was all swollen. She was in shock, praying for God to take her. There was little we could do to help her, try as we did. We just tried to keep her warm, cover her with blankets. She was coughing, screaming, and praying for God to let her die." The children are now afraid to go in that room. They call it the "booger room."

Many others who reached high ground wandered aimlessly that day and into the night, building makeshift shelters, often huddling with scores of survivors in ramshackle, unheated wooden huts, or just freezing out in the cold next to fires. Sam Baisden, a retired coal miner, remembers the people on the hillside: "There was fires built and things. There was a whole bunch of people there. I don't know how many was on the hill. I really don't know. It was so many of them you couldn't count them if you wanted to. Kids and women and men—families.

"One old colored man, he was setting there shaking. I carried two railroad ties up and put them on the fire for him. He died right after that."

Throughout Saturday and into Sunday, hundreds upon hundreds of homeless survivors wandered aimlessly, in a state of near if not actual shock, into the homes of friends or into Man at the mouth of the Valley itself. Even by Sunday afternoon it was still virtually impossible for separated family members to know whether their loved ones had perished or survived.

Julian Scalf tells the story of his day-long search for his uncle and aunt. The National Guard required that he first "seek his family out" in the hospital and the morgue. "All four of us went through the morgue . . . twenty-six bodies were there at that time in every twisted position, just like they died. Babies, youngsters, old people, all ages . . . black with muck . . . the faces of my brother and the other two in our party were as white as sheets when we got outside, and I suppose mine was too. And just as we left the morgue they brought in a girl of about thirteen, with her head almost cut off."

The National Guard then issued him a pass, and he agreed to report back to them the names of all survivors in one part of the Valley. The five-mile drive up the Valley took over two hours and Mr. Scalf noted with horror at the time: "You can't imagine, houses smashed, trailers destroyed, piles of trash and logs and brush, bodies at almost every bridge abutment. National guardsmen were pulling bodies out of wrecked houses, out of the black mud and piles of driftwood. The dead were strung all along the creek and roadside. They were loading bodies into helicopters and trucks. When we got to about a mile below Riley, there was a dead man sitting in the door of a ruined trailer."

Luckily, Mr. Scalf soon found his aunt and uncle. On his way back to the mouth of the Valley, he saw many small fires all along the hillside where "the ones that were alive were wet and were trying to stay warm." He brought the first news of fifty survivors of the coal camp of Riley.

The search for the dead continued with literally the entire community involved. Some will never know what happened to their loved ones. The bodies of three young children never were identified. They now lie in a grave marked with this stone:

THIS — TINY — TRIO
Boy — Girl — Boy

Who were the Victims of the Feb. 26, 1972 Buffalo Creek Disaster are unknown to us by name but to Our Heavenly Father they are known as three little angels.

VI

“They Wasn’t for Sale”

By the beginning of the summer, the federal government had provided trailers for the survivors to live in, rent-free, in camps hastily put up on whatever flat land the government could find in the area. At first these trailers seemed like “manna from Heaven.” The people could at least get their families together again under one roof, although that was true only for the smaller families. The larger families had to split up because the trailers were so small—with the husband and boys sleeping in one trailer, the wife and girls sleeping in another.

But the trailer camps soon felt like concentration camps to the distraught survivors. The trailers were not assigned so that people from a particular coal camp would once again be living next to those from their same coal camp. As one survivor complained, “There isn’t one family in our trailer park that we were really close friends with, and so we feel like we’re in a strange land even though it’s just a few miles up Buffalo Creek from where we were.”

As the summer sun beat down on the trailers, the people soon looked upon them as the new enemy. They said that during the day they baked in the hot trailers, which they soon came to call "hot boxes." And at night, when the rains came, reverberating like the death beat of drums on the flat tin roofs, they told us they paced the floors, feeling like "condemned prisoners."

While they suffered in these trailer camps, we pressed forward with the legal research we had to complete before we could begin their lawsuit for them. Many legal questions had to be asked and answered before we could file the complaint which would signal the beginning of this lawsuit.

The most obvious question was, Whom do we sue, Buffalo Mining Company or Pittston? We could easily file a complaint against the Buffalo Mining Company in a state court in Logan County, where, the local prosecutors had assured me, lawyers could pull some tricks they could not get away with in federal court. But I was afraid those tricks would be pulled on me, rather than by me, if I tried to fight the coal company on its home turf. So we tried to find a way to get the case into federal court, where we felt more at home.

Since Washington is our home territory, we thought of suing there. This was possible if we could claim that the failure of the dam was caused by Pittston's violation of federal law. We also considered suing Pittston in New York, its home base, thinking that Pittston might have less control over juries in New York than it would in West Virginia. But our legal research convinced us Pittston could always force a transfer of the case to West Virginia from Washington or New York.

So we settled on trying to sue in the federal court in West Virginia—the United States District Court for the Southern District of West Virginia, to be precise. But we also had to decide where to file within the Southern District. Almost every federal district now has more than one federal judge. The cases are distributed to them in rotation as they are filed. So it is rare that you can file your case in such a way as to pick the judge who will try it. But a limited form of judge-shopping was possible in the Southern District of West Virginia. Huntington, West Virginia, had only one federal judge, Judge Sidney L. Christie, while Charleston, West Virginia, had two, Judge Dennis R. Knapp and Judge K. K. Hall. If we filed in Huntington, we would be certain that Judge Christie would preside over our case. But if we filed in Charleston, either judge there might end up with our case.

Judge Christie had been a federal judge for eight years, so we began collecting and reading his opinions. One of the first things lawyers often do is find out what the judge has said in similar cases in the past. This is time-consuming legal research, since judges' opinions are referenced by the names of the cases rather than the particular judge's name.

I first ran into this problem soon after I graduated from law school. I had been called down to the Justice Department on the Sunday of the riots at Ole Miss, when the department was helping James Meredith become Ole Miss's first black student. President Kennedy was planning to go on nationwide TV in two hours to make a speech asking the citizens of Mississippi to abide by the United States Constitution and to avoid violence. He intended to include in his speech some reference to a famous Supreme Court Justice from Mississippi, Lucius Quintius Cincinnatus Lamar. President Kennedy knew quite a bit about L. Q. C. Lamar. He had devoted one chapter in his Pulitzer Prize-winning book, *Profiles in Courage*, to Lamar, a famous Southerner who had done much after the Civil War to try to reconcile the differences between the South and the North. So one of President Kennedy's speech writers asked us to read all of Lamar's Supreme Court opinions to find some language which President Kennedy could use in his speech. To find those opinions, I sat in the empty fifth-floor Justice Department library that Sunday, laboriously thumbing through the pages of the Supreme Court reports to find cases in which

Lamar had written the opinion. Unhappily, it took me too long to come up with these cases and any significant language, although the President still invoked the memory of Lamar's example in his speech.

The law clerks assigned to find and read Judge Christie's opinions had more time, and eventually they came up with the fact that Judge Christie had once written an opinion permitting an injured person to sue a corporation's sole shareholder for injuries caused by the corporation. This is known as "piercing the corporate veil." Rarely is such a suit allowed. The general rule, as corporate law developed in England and as it has been followed here in the United States, has protected shareholders from lawsuits when the corporation they own common stock in causes damage to others. Thus, if General Motors harms you, you can sue GM, but you can't sue Joe Jones who owns fifty shares of GM common stock.

To get into federal court, our West Virginia plaintiffs would have to sue the out-of-state Pittston Company. But to sue Pittston, the Buffalo Mining Company's sole shareholder, they would have to pierce the Buffalo Mining Company's corporate veil. Thus, it would be critical to have a judge, like Judge Christie, who understood the necessity for piercing a corporation's veil in certain cases.

Since piercing the corporate veil is so rarely permitted, however, we tried to find some other way to sue Pittston without relying only on the diversity of citizenship between Pittston and the West Virginia plaintiffs. If a lawsuit raises a federal question, the suit may be brought in federal court even without diversity. So we considered making this a federal question by arguing that Pittston had polluted a navigable stream in violation of the federal Navigable Streams Act. But this was unnecessarily complicating an already complex case. Pittston was not at fault because of some technical violation of a federal statute.

So we decided to gamble, to rely only on diversity

and to sue only Pittston, hoping we could convince the court of the need to pierce the Buffalo Mining Company's corporate veil. I prayed that when the time came for the court to make that decision, we would be able to detail enough facts about Pittston's responsibility for this disaster to persuade the court not to leave hundreds of victims without any federal remedy.

We also spent a substantial amount of time researching West Virginia law. Even if we could keep the case in federal court, the judge still would be required to apply West Virginia law to the facts of the case. I was outraged when I first learned that West Virginia had a $110,000 maximum recovery for those killed by others. Many states have much higher limits or no limits at all. For example, New York has no statutory limit. Thus, if a resident of New York dies in an airplane crash, his life may be worth $2 million to his family, whereas a West Virginia resident who dies in the same crash can have a maximum value to his family of only $110,000. I wanted to find some way to get around West Virginia's wrongful death limits.

I sent a law clerk to Charleston to research the background of the West Virginia statute. Legal research sounds like a dull process. Someone goes to a law library and reads cases from old musty law books. But legal research can be much more exciting than that. For this problem, for example, the law clerk had to talk to the legislators, at least those who were still around, to ask why they wrote the bill as they did. His research was full and complete, but fruitless. We could find no way to attack the legality of the limits.

We also read West Virginia cases to determine the standard for recovering "punitive damages" against Pittston. In cases where the defendant's conduct is more than merely careless or negligent—where the conduct is willful, or wanton, or reckless—the plaintiff can recover not only "compensatory damages," to pay him back for his own loss, but also "punitive damages," to punish the defendant and deter him and others from ever again harming the plaintiff or others.

If the plaintiffs could prove only that Pittston's conduct was negligent, they'd collect only compensatory damages. In that case, it might be better for them to take the money Pittston now might offer at its claims office, rather than wait many years, without the use of that money, and then have to pay our lawyers' fees out of the amount they finally recovered. On the other hand, if we could prove Pittston's conduct was reckless, and collect punitive damages as well as compensatory damages, it might be better for the people to sue than settle at the claims office. This was a real dilemma for the plaintiffs, and some who had signed retainer letters with us soon decided they could not wait and gamble on a larger verdict later. They went ahead and accepted Pittston's offer.

Luckily, though, many of the plaintiffs did not have such pressing needs for money. In fact, there was little immediately to spend the money on. A new wider highway was planned for the Valley, and until the highway route was selected, the people could not build homes back in the Valley. So, many of the plaintiffs stayed on, rent-free, in the government trailers. There was also money available from the Small Business Administration, which made loans quickly available to the people. Without this federal help of homes and money, many, if not all, of our plaintiffs might have been forced to reach immediate settlement with Pittston at the price Pittston was then willing to pay.

Pittston also helped by making its settlements unreasonably low. The very first settlement, which Pittston announced with great public fanfare, amounted to a total of $4,000 to a Pittston employee who had lost a six-room house. Thus, although it was now clear that Pittston would pay money to those who filed claims with it, it was evident that not much money would be paid. All during the summer months when I was interviewing our clients at Charlie's gas station, I heard stories of the ill treatment they felt they were receiving at the claims offices and the difficulties they were having in obtaining the numerous receipts for lost items which Pittston was insisting upon. Needless to say, putting together a proven list of items from re-created receipts invariably caused the people to forget items. And even when they finally came up with their total list, Pittston would offer them only about half of what they were asking for.

In various public statements Pittston also began to argue that even if Pittston, not God, was at fault, and even if Pittston, not the Buffalo Mining Company, was the proper company to sue, Pittston's conduct was not negligent, and certainly not reckless. This is the line which Pittston's president, Nicholas T. Camicia, took in his testimony at Senate hearings three months after the Buffalo Creek disaster: "The embankment had been constructed by experienced coal-mining men in accordance with methods and techniques for years characteristic of the manner in which such water impoundments have been constructed throughout West Virginia and elsewhere."

But this "custom and usage" argument was not a safe refuge for Pittston. During our research to determine the legal test for reckless conduct, as opposed to merely negligent conduct, one of our summer law clerks came across a case decided in 1926 by the United States District Court for the Southern District of West Virginia. In this case, which had striking similarities to the Buffalo Creek disaster, a refuse pile which blocked a stream gave way during the rainy season. The refuse pile was burning, as are many of the refuse piles in West Virginia. Water caused this burning refuse pile to explode and then rush down a mountain hollow where it buried a home and its seven occupants. The coal company's defense was "act of God," and anyway, they argued, the coal company was merely following normal custom and usage in building its refuse pile across the stream. The federal district court denied the act-of-God claim and also held that custom and usage was no defense. The court said that if every coal company is doing something wrong, then every coal company is at fault. It is no defense for one of them to say that the others are just as careless.

After the jury returned a verdict against the coal company, the company appealed. The United States Court of Appeals for the Fourth Circuit, the court between the federal district court level and the Supreme Court level, upheld the federal district court in the following language:

> It seems clear that the piling of the waste material so as to obstruct a natural watercourse, impounding the water behind the obstruction, and allowing the waste pile to burn for a number of years, constituted negligence.

What a case. If the coal company hadn't appealed we might never have found it. District court opinions are rarely published. But all the opinions at the appellate level are published. Also, a district judge's opinion, even if discovered, is not binding on another district judge. But an appellate decision is binding on all the district courts in the jurisdiction of that court of appeals. So this opinion of the Fourth Circuit Court of Appeals would be binding on a federal district judge in the Southern District of West Virginia—Buffalo Creek's district—if we could stay in federal court.

In the law it is very rare to find a prior case exactly like yours. You always search for such a case. But your opponent always finds some way to distinguish any case you find. If the railroad kills your cow at the crossing, and you find a case where the court required this same railroad company to pay when it killed another cow at this same crossing, in the same way, your opponent will say that other cow was a different color. But we found the cow of the same color—a refuse-pile disaster, in West Virginia, decided by a federal district court in the Southern District of West Virginia and upheld by the Court of Appeals for the Fourth District. At the least, we could defeat any act-of-God claim or custom-and-usage claim—and thus recover for negligence. But the earlier case didn't say the coal company was reckless, only negligent.

We got further support for the drafting of our complaint in various reports which now began to filter out of the federal government on the causes of the Buffalo Creek disaster. One such report, by William Davies, became critically important to us. Mr. Davies had studied a number of refuse piles in this country after a tragic refuse-pile disaster occurred in Aberfan, Wales.

The Aberfan disaster occurred almost six years before the Buffalo Creek disaster. At 9:15 A.M. on October 21, 1966, at Aberfan, Wales, a massive coal-waste refuse pile (or spoil heap, as such refuse piles are called in Great Britain), loosened by heavy rains, slid down the mountainside, quickly covered two nearby houses, crossed a canal, surmounted a railway embankment, and then engulfed and destroyed a school and eighteen houses. In all, the refuse pile moved rapidly over 2,000 feet, killing 144 people. One hundred and sixteen were children, most of them between the ages of seven and ten.

The Aberfan disaster received immediate worldwide attention, primarily because so many helpless schoolchildren were killed. Walter Cronkite began the CBS-TV nightly news on October 21, 1966, with this report:

> Aberfan is a coal-mining village in Wales. When mining first came to Wales, it brought jobs; it also brought ugliness and fear. Ugliness in the great dark hills of waste from the mines; fear that at any moment and without warning, those hills would give way and lives would be snuffed out. Today, in Aberfan, fear became a stark reality.

The report directly from Aberfan, with film, then followed:

> The biggest rescue operation since World War II is going on here in Aberfan, but the full horror of the disaster is becoming more clear by the hour. . . .
>
> The tragedy began shortly after nine o'clock this morning, just after the students had taken their seats in the classrooms. Torrential rains had weakened the huge mountains of slag above this little mining village, and suddenly the entire mountain slid down like a soft, mucky glacier, a massive landslide, 40,000 tons of it, burying the school, a row of miners' houses, even a farmhouse.

CBS continued its filmed coverage of Aberfan on October 22 and 23 and again on October 29, 1966.

The *Charleston Gazette*, *Charleston Daily Mail*, *New York Times*, and *Washington Post* all carried the report of the Aberfan disaster on their front pages on October 22, 1966. *Time* and *Life* also carried the Aberfan story in a number of issues.

Surely after Aberfan no one in the coal industry could argue that coal-waste refuse piles are not extremely hazardous. Indeed, the refuse pile at Aberfan was not as hazardous as the refuse-pile dams on Middle Fork at Buffalo Creek. Aberfan's pile did not block a stream or watercourse—although it was improperly located over an underground stream—and did not impound black plant water. But it was a massive killer and again emphasized the importance of ensuring that water is kept away from coal-waste refuse piles.

The Aberfan disaster triggered investigations in this country into similar refuse piles, and Mr. Davies had been deputized by the U.S. Geological Survey to make those studies. When he visited Buffalo Creek in 1966, there was only one dam across Middle Fork. He determined that this dam, Dam 1, was stable, but "subject to large washout on north side from overflow of lake." Soon after his report, Dam 1 did overflow, in March 1967.

After the Buffalo Creek disaster he was called on again by the federal government, this time to determine the reason for the failure of Dam 3, one of the dams on Middle Fork which had been built after his earlier report. Mr. Davies's report on the Buffalo Creek disaster formed the basis for the allegations in our complaint as to the ways in which Pittston's conduct was reckless in the construction and maintenance of the dam which caused the Buffalo Creek disaster.

There was a similar report prepared by Garth Fuquay for the Senate hearings on the Buffalo Creek disaster. Mr. Fuquay is chief of the Engineering Branch of the Los Angeles Division of the U.S. Army Corps of Engineers. He is an expert in soil mechanics and has been involved in the construction of numerous earth-filled dams all over the world. His conclusions were similar to those of Mr. Davies. With these two reports, and the lessons which Pittston should have learned from the Aberfan disaster, we had a solid basis for alleging in our complaint that the dam itself was at fault, and not God.

We had to find a West Virginia lawyer to join us in signing the complaint and in continuing the litigation. Even though I was admitted to practice before the federal courts in Washington, D.C., and could be admitted, on a simple motion, to try this case in the West Virginia federal court, it is customary to have a local lawyer sign the complaint and other papers filed with the federal court in his jurisdiction. A local lawyer can also help with the nuances in the local federal practice—such as getting extensions of time to file papers, and working with the local lawyers on the other side in making certain that various steps in the proceedings are carried on in a courteous and lawyerlike fashion. It is also helpful to find a lawyer who personally knows the judge's idiosyncrasies and predilections. For example, some judges decide motions on the written briefs and pay no attention to oral argument. Other judges never read the written material and rely instead on the oral argument.

In this case, finding a local West Virginia lawyer raised some difficult problems. The survivors had come to us, in Washington, D.C., because they felt there was no West Virginia lawyer they could rely on. If we could find such a lawyer, we would be telling our plaintiffs, to some extent, that they were wrong in coming to us. And anyway, we doubted we could find such a lawyer. Nevertheless, we had to find some local lawyer to join with us in the filing of our complaint.

Harry Huge called a friend in the law firm of Steptoe & Johnson in Washington, D.C. Steptoe & Johnson is an unusual firm. It has offices in Clarksburg and in Charleston, West Virginia, as well as in Washington, D.C. The firm began in Clarksburg many years ago when Louis Johnson was still a West Virginian. Johnson later became the Secretary of Defense in the Truman Administration, and after he left that Administration, he stayed on in Washington to head the office of Steptoe & Johnson. Harry hoped Steptoe & Johnson could recommend a West Virginia lawyer for us. Of course, he did not expect Steptoe & Johnson, one of the largest law firms in West Virginia, to offer to join us in the lawsuit, since they had many coal-company clients.

To Harry's great surprise, his friend suggested a Steptoe & Johnson lawyer in the Clarksburg office who might be interested in helping us. This lawyer, Willis O. Shay, known to everyone as "Bud" Shay, has a great reputation as a West Virginia trial lawyer. He had just won one of the largest jury verdicts in the history of the state, and he had shown his own concern for public interest matters by bringing a number of environmental lawsuits. We jumped at the chance when Bud agreed to join us. Since Bud's recent victory had been in a case tried before Judge Christie, getting Bud Shay on board was a great coup.

There was one other major area of legal research to be completed before we filed our lawsuit. This was the question of recovery for the survivors' mental suffering. In a typical automobile accident, if you are physically injured, you can recover for your physical damages, such as medical bills for fixing your broken leg. You may also recover for any pain and mental suffering in connection with that physical injury. In the Buffalo Creek disaster, though, most of the survivors suffered no physical injuries. They had escaped to the hills before the water came down the Valley. So Pittston apparently assumed there could be no claims for pain or mental suffering.

But Roland Staten's story forced me to think about the severe mental suffering of the survivors of this disaster. When I first met Roland, and heard him recall his vivid memories of his wife's last words to him, to save their baby, which he then was unable to do, and saw his anguish as he told of having his twenty-two-month-old son washed from his arms, I became emotionally upset. At the time I met Roland, my own son was almost twenty-two months old, and the thought of being unable to save my son, of living with that kind of guilt, was difficult for me to deal with. Roland himself told me how troubled he was. He said he immediately went back to work in the mines to keep from thinking about the disaster day and night. And he began working extra long hours just to fill up his waking day so he would be physically tired enough to sleep without nightmares.

I did not really understand what was going on in Roland's mind or in the minds of the other survivors. But I had a feeling their nightmares, their fears of rain and water, which I heard constantly repeated, were a kind of injury, mental suffering or mental anguish, more pervasive and more serious even than the loss of their homes and all their possessions.

Near the end of the summer, I gave a luncheon talk to the lawyers at Arnold & Porter about this case and about my plans for the complaint. I said the complaint would allege damages for some kind of mental suffering, even though I did not fully understand what that suffering was all about.

One of my partners had spent a considerable amount of time dealing with psychiatrists in a landmark "insanity defense" case, the *Durham* case before the United States Court of Appeals for the District of Columbia Circuit. He suggested I contact Dr. Robert J. Lifton, who had written books on the survivors of Hiroshima and the survivors of the Nazi concentration camps. Dr. Lifton had identified a survival syndrome, a kind of survival guilt, that the survivors of disasters often feel. This guilt causes the mental suffering so often seen in survivors. I decided to contact Dr. Lifton, but I did not have time to do so before filing the complaint.

The complaint as filed did contain a claim for damages for mental suffering, but there was no discussion of the survival syndrome as a particular element of damages. Instead, we called this mental suffering "psychic impairment" and claimed that each plaintiff, the children as well as the parents, could recover damages for it. By listing the children separately, we also increased the number of plaintiffs from approximately 200 to approximately 450. I was trying to make it appear that we had a much larger and more substantial case against Pittston than merely 200 adults or only 100 families.

A complaint must state the amount of damages claimed. In this case, that was an overwhelming and almost impossible task. Even by the end of the summer, we still did not know the dollar value of each of the thousands of items the plaintiffs had lost.

Later Pittston would question the plaintiffs at length about the "market value" of their possessions. For example, they asked sixty-four-year-old Rex Howard, "Do you have any opinion as to the market value of the things that were washed away, such as the coal, the coalhouse, the walk, the tools on the front porch, the shrubbery, and the shade tree, as of February 26, 1972; that is, what a willing buyer would have paid a willing seller for those various articles, taking into consideration their age, the character or description of the item, and their condition?"

Mr. Howard's response put Pittston's lawyers in their place. "I have no idea, for they wasn't for sale and I don't know what a tree twenty years old would cost you. I haven't got time to grow another one."

The problem was that these items were valuable to the people in terms other than market value. For example, what would it take to repay you for seeing your whole world destroyed, to lose all your roots, the only photos of your children—when they were little, or of your son who died in Vietnam, or of your parents? To lose your family Bible, the wedding dress your mother wore and you wore, and which you promised your daughter she could wear? The penny minted the year your father was born, given to him by his father, which you promised your son you'd leave to him?

Eventually we claimed that the plantiffs' real property and personal property losses totaled $11 million, a figure we made up.

We also had difficulty trying to estimate a dollar value for mental suffering. We had no precedent to guide us. At that point, no one had ever attempted to collect damages for survivors, merely because they survived. We knew very little about the underlying psychiatric bases for the survival-syndrome claim. I asked one of my partners to suggest a dollar amount of damages for the mental suffering I felt to be endemic to the survivors. The largest number he could think of for this psychiatric claim for each survivor, bearing in mind that most of the survivors were not even in the water, was $25,000. So I doubled this and alleged $50,000 of damages for each plaintiff's psychic impairment. This meant that our claim for mental suffering, for the 400 or so plaintiffs suing for their own personal injuries, totaled approximately $20 million.

Finally, we made a claim for punitive damages of $21 million, half of Pittston's $42 million net income for the past year. Since the disaster was a significant event which might affect Pittston's net income, Pittston had to report to the Securities and Exchange Commission on the possible financial effects of the disaster. In its report, Pittston repeated the assurance it gave to its own stockholders—the disaster would not have any "material effect on its consolidated financial position." I was appalled at the callousness of that sweeping assurance. Even with its immensely formidable "consolidated financial position," how could Pittston be so sure?

I used this statement against Pittston. Since Pittston had said the disaster would have no material effect on it, and since the amount of punitive damages is supposed to be large enough to punish and deter the wrongdoer, I alleged in the complaint that $21 million in punitive damages was necessary to ensure that this disaster *would* have a material effect on Pittston. We were now ready to file the complaint.

From the time the survivors first contacted us, we had hoped Pittston's lawyers would call us, to meet with us, maybe even to offer a fair settlement of our clients' claims. But they never called. Now that we were ready to go to court, we debated whether we should make this first call ourselves, to initiate settlement discussions. But we decided that would look weak. We had to plow ahead, file the complaint, begin discovery, show Pittston we would go all the way, even to trial if necessary. Only then, from a position of strength, could we get the best possible settlement.

VII

"Flagrant Disregard"

The day before we filed the complaint we heard startling news. A commission created by the governor of West Virginia—the West Virginia Ad Hoc Commission of Inquiry into the Buffalo Creek Flood—had just issued its report. The governor released some of its findings to the press. In particular, the commission found:

> The Pittston Company, through its officials, has shown flagrant disregard for the safety of residents of Buffalo Creek and other persons who live near coal-refuse impoundments. This attitude appears to be prevalent throughout much of the coal industry.

A ringing indictment of Pittston, and the whole coal industry. Already our decision not to begin settlement discussions had paid off. With this report we could insist on punitive damages for Pittston's "flagrant disregard."

I had fully expected the report would be a whitewash, as

did everyone else. Indeed, a group of West Virginians formed their own commission to investigate the Buffalo Creek disaster because they feared the report of the governor's Ad Hoc Commission would just be a cover-up. The *Charleston Gazette* agreed:

> At the time the commission appointments were made, this newspaper voiced its suspicion that a whitewash was in the offing, basing the prediction on the makeup of the commission. It included representatives of federal and state agencies responsible for regulation of coal refuse impoundments. It included members known to have ties to the coal industry.

Dr. Jay Hillary Kelley, the chairman of the governor's Ad Hoc Commission, is the dean of the West Virginia University School of Mines. Thus, he is considered a coal-industry spokesman. During the hearings, he appeared to protect Mr. Camicia, Pittston's president, after another commissioner tried to get Mr. Camicia to admit Pittston's responsibility for the Buffalo Creek disaster.

> Q [by Commissioner Hylton]: My point is that after our report is filed, there will be a lot of searching and a lot of hair splitting and a lot of nit picking to try to fix a responsibility, and if we could get you to say your company was responsible for the disaster, it would save a lot of trouble.
>
> Mr. Staker [Pittston's legal counsel]: If the Commission please, at this juncture I would like an opportunity to confer with Mr. Camicia, before he makes a further utterance.
>
> Chairman Kelley: Sustained. I think maybe you could withhold that question.
>
> Commissioner Hylton: All right.
>
> Chairman Kelley: Let me add something to this, Commissioner Hylton. Even if this witness would assume responsibility, I don't think it would change this Commission-

> er's point of view, and other Commissioners' points of view on the causes and effects and such.

Thus, the commission's findings that Pittston had "shown flagrant disregard" came as quite a surprise.

I immediately went to the governor's office to get a copy of the report. They refused to give it to me. They said it was not a public document. The governor would not let it out of his hands. Apparently, he feared some unscrupulous politician might use it to accuse the state, or even the governor, of complicity in the disaster. With West Virginia's gubernatorial election now less than two months away, the governor, who was running for reelection, wasn't taking any chances. I argued and argued, to no avail. I left. But then I began thinking, and got angrier and angrier. No one was more entitled to that report, or could make better use of it, than I, as the people's lawyer. I turned around and went back to the governor's office.

I wouldn't leave. I was shunted from one secretary to the next. This one finally said I could look at it, but not take notes. That one said I could take notes, but only for a half-hour. Someone else finally said I could take my time reading it there. In the end, after several hushed phone calls, and some meetings behind closed doors, a copy was found for me to take with me. But I couldn't see any of the exhibits or documents supporting the report. For now, though, the report was enough.

We filed our lawsuit one day after the newspapers reported the commission's conclusions. The newspapers headlined our $52 million lawsuit and referred back to the commission's report that the Pittston Company had shown "flagrant disregard" for the lives of the people of Buffalo Creek. So much of life is luck. Maybe Pittston would think we had been readying our lawsuit hand-in-glove with the preparation of the governor's Ad Hoc Commission report.

The commission called for the convening of a special

grand jury to determine whether criminal indictments should be brought against anyone as a result of the Buffalo Creek disaster. This became a hot political issue, especially after a newspaper reported the fact that the Logan County prosecuting attorneys had accepted a number of civil suits on behalf of plaintiffs who were suing the coal company, and therefore were in no position to conduct the grand jury proceedings. So, the Logan County judge appointed two special prosecuting attorneys, Dean Willard Lorensen of the West Virginia University College of Law, and Lafe Chafin, a Huntington lawyer and former prosecuting attorney, to prepare the case for the Special Logan County Grand Jury. The judge then announced that the grand jury would not be called until after the November gubernatorial election.

The grand jury met soon after Governor Moore's reelection. After hearing testimony for only a few days, it decided that no criminal indictments should be brought against anyone, not even against Pittston, for the death of the 125 people from the Buffalo Creek Valley. Its public finding was:

> That the Special Grand Jury in and for the body of the County of Logan, West Virginia, having considered evidence presented to it regarding the disaster on Buffalo Creek on February 26, 1972, and after careful deliberation finds that it will return no true bills with regard to criminal charges growing from that tragedy.

A grand jury's deliberations are secret, so I never did learn how the jurors decided that no one, not even Pittston, was criminally responsible for so many deaths. Many of the people in the Valley assumed the grand jury didn't indict Pittston because of coal-company pressures. I assumed the same thing and prayed we could keep our lawsuit in federal court, far away from Logan County's jurors. I later learned, though, that it may have been the special prosecutors, worried about the "corporate veil" question, who made it possible for Pittston to escape criminal indictment. Lafe Chafin told me that he let the grand jury know it was "legally and particularly practically difficult to sustain an indictment against Pittston," because they couldn't pierce the Buffalo Mining Company's corporate veil. Lafe Chafin felt an indictment against Steve Dasovich, the man in charge of the Buffalo Mining Company operations, was about as far as the grand jury could go.

The grand jury, on the other hand, according to Lafe Chafin and Dean Lorensen, felt it would be capricious to fix all the responsibility on Mr. Dasovich. If they couldn't indict Pittston and all the officers up and down the line, then they'd indict no one. Still, it was a "mighty close vote."

I was happy to learn that the grand jury wanted to go after Pittston, or no one. I was even happier that it decided not to make Mr. Dasovich the scapegoat. Although I was as eager as anyone to have the special grand jury bring indictments, the possibility of a criminal indictment of Mr. Dasovich raised difficult problems for me as a lawyer representing the survivors in a civil action. If the grand jury indicted him, I was afraid he then would refuse to testify in our civil case, or seek a delay in our civil case, until the criminal proceedings were completed. And singling him out, instead of Pittston, would have been adverse publicity for our case. So the grand jury's decision was not too unfortunate for us, although it once again highlighted our corporate-veil problem.

VIII

"Mere Puff and Blow"

It was now up to us, and our civil suit. The filing of our complaint with the clerk's office started the chess match. The clerk gave a copy of the complaint to the federal marshal. He served it on Pittston. Pittston now had twenty days to answer our complaint, to admit, or deny, the plaintiffs' allegations of wrongdoing. But a defendant rarely answers within twenty days.

We expected Pittston to ask for innumerable delays before answering our complaint, so we took the offensive without waiting to hear from them. We filed a notice to take the oral deposition of Pittston's president and vice-president one month later. A deposition is similar to testimony in open court, with lawyers from each side present and a court reporter there to take down the testimony under oath. However, a deposition usually is taken in a lawyer's office and the judge is not present. Moreover, deposition testimony may be used later in court only in certain special circumstances. Depositions are supposed to help each side learn all the facts, under oath, about the other side's case. An important assumption is that full discovery of all the facts, prior to trial, will permit the parties to evaluate their cases properly, and then reach settlement, without wasting the court's time with a trial.

When I wrote Pittston that I wanted to depose its president and vice-president I also attached an appendix categorizing the kinds of documents I wanted them to produce at their depositions. This document request was based in large part upon the findings of the governor's Ad Hoc Commission report and on the findings and transcript of the Senate hearings at which Pittston's officials had testified. The Senate committee made a large number of critical documents available to us by publishing almost every document it obtained in its investigation. Once the documents were published, they were available to us and to everyone else.

As we assumed, almost immediately after Pittston's lawyer received the complaint, he called to ask for a sixty-day extension to answer. We agreed to a thirty-day extension. We also agreed to meet with him in Washington, D.C., two weeks prior to the time when Pittston was to file its answer and ten days before the depositions were supposed to commence. I missed this first meeting. I had ruptured a disc in my back and had to have an operation. An associate captured its flavor for me.

Pittston's main lawyer at this first meeting was Zane Grey Staker, a legendary trial lawyer from Kermit, West Virginia. Dan Murdock, from Pittston's New York law firm of Donovan Leisure Newton & Irvine, also was there. Mr. Staker made it clear he was in charge of Pittston's side of the case. He'd been Buffalo Mining Company's legal counsel for many years, even before Pittston became its sole stockholder. He'd made a "handsome living" as Buffalo Mining Company's lawyer, at least for a "country lawyer" like himself. He assured us that the conduct of litigation in West Virginia was more relaxed than in the "more sophisticated areas of the country" and hoped we'd "take cognizance" of that fact. He didn't contemplate running into court for rulings on minor issues and hoped a "harmonious climate" could be established in which we'd always "talk things over." We said we too hoped a cordial relationship would prevail, but emphasized this could come about only if the proceedings moved along expeditiously.

Mr. Staker then brought up settlement. He spoke in his capacity as general counsel of Buffalo Mining Company, constantly reminding us that Pittston was not an appropriate defendant. He added that the Buffalo Mining Company had "sought good will" through the opening of three claims offices. They were there "to meet a human need." The settlements had been made on a "generous basis." If we knew the relevant facts, we too would certainly agree the settlements were "handsome" ones. "Buffalo Mining" had now settled with almost all the claimants, except our 400 or so plaintiffs. Buffalo Mining would do "whatever" he said, and he was sincerely interested in pursuing the possibility of settlement. We showed no interest in settlement.

Mr. Staker repeated his "profound persuasion" that we were suing the wrong people in suing Pittston. As general counsel of Buffalo Mining Company, he never regarded Pittston and Buffalo "in the same bed." Anyway, he said, Buffalo Mining Company had enough money to take care of all claims. But when we asked how much insurance coverage, if any, they had, he said he was not at liberty to disclose that information to us.

He then agreed to begin producing documents, but only on the corporate-veil question. He wanted the issue of whether we could sue Buffalo Mining Company's stockholder resolved at the outset. I didn't blame him. If they could get out of federal court on that issue, we'd be forced into state court and there wouldn't be much left to our case. We had no choice but to go along, for the time being, with this narrowing of the case to the corporate-veil question, if only to get our hands on some of the documents we'd asked for. We also agreed to postpone the beginning of depositions for a month and told Pittston we had additional plaintiffs who would be added to the case.

Soon thereafter, Pittston formally moved to dismiss our case—on the ground that Pittston was "not a proper party to this action." Its brief said that Pittston "operates in large measure as a holding company." Pittston's oil division includes a number of oil companies such as Metropolitan Petroleum Company in New York and New England. Its trucking and warehousing division is primarily its wholly owned United States Trucking Corporation. And it owns most of the stock in Brink's, the armored-car service. Finally, it is the sole stockholder in a number of coal companies in Virginia, Kentucky, and West Virginia, which it operates as separate subsidiary companies within the Pittston Coal Group.

Pittston added that it didn't own the dam at Buffalo Creek. That was owned by one of its subsidiaries, the Buffalo Mining Company, and Pittston was merely a stockholder in the Buffalo Mining Company. In fact, Pittston said it had only recently purchased all the Buffalo Mining Company's stock for $7 million, in June 1970, less than two years before the dam gave way. It also pointed out that Buffalo Mining Company had its own duly elected president, Irvin Spotte, and its own duly elected vice-president, Steve Dasovich. And it held formal stockholder meetings and formal meetings of its board of directors, as separate, independent corporations are supposed to do.

Pittston might have said, but didn't, "You don't sue Pittston when a Brink's truck runs you down, you sue Brink's. So you should sue Buffalo Mining Company, not Pittston, when its dam destroys your home." Pittston did say that courts rarely permit a plaintiff to pierce the corporate veil, unless the use of the corporate entity is fraudulent or illegal in some way. As Pittston's brief pointed out:

> [A] plaintiff seeking to persuade a court to ignore separate corporate entities must show not only an excessive degree of control over the subsidiary, but the purposeful exercise of that improper dominance to commit, behind the shield of corporate limited liability, a fraudulent, illegal, or otherwise wrongful act directly productive of the injury or loss on which the plaintiff sues.
>
> * * *
>
> The short of the matter is that because plaintiffs regard Pittston as a more desirable defendant than Buffalo Mining for their purposes here, they seek in a manner wholly defiant of the facts to fabricate a wrongful domination of Buffalo Mining by Pittston that has never in truth existed, and without which, in light of the principles of law treated with hereinabove, vicarious liability for the acts of Buffalo Mining cannot be laid to the door of Pittston.

We had a month to respond. Our entire legal strategy was riding on this narrow, technical procedural question.

We didn't agree with Pittston that "fraud or illegality" was required to pierce the corporate veil. But we did have to show that the Buffalo Mining Company wasn't, in fact, operated as a separate, independent corporation.

We now had the transcripts of the sworn testimony of Pittston and Buffalo Mining Company employees before the governor's Ad Hoc Commission. The governor had refused to give us access to these transcripts. However, we learned that Pittston had purchased copies during the hearings. This was fortunate for us. Because of our document request, and because these transcripts were now in Pittston's possession, Pittston had to give copies to us. These transcripts, along with documents from the Senate hearings, from the public files of the United States Bureau of Mines, and from Pittston's New York files, enabled us to make a fairly good argument for piercing the Buffalo Mining Company's corporate veil. But we were still handicapped. Pittston refused to give us any documents from Buffalo Mining Company's files, and we hadn't yet deposed any of Pittston's officers.

Nevertheless, we tried to prove, with what we had, that Pittston was the proper defendant. To a nonlawyer, our arguments will appear irrelevant and far removed from the real issues of the disaster. Dickens is quoted often for writing, "The law is a ass." And one of my clients has this framed quote from Shakespeare's *Henry VI*, Part II, act 4, scene 2, on his wall: "The first thing we do, let's kill all the lawyers." It is true that lawyers' arguments never seem to get to the point. But sometimes that is the point. Pittston was unwilling to meet us on the merits of the case. So they argued procedural questions instead. In law school I had a professor who maintained, "I can win any argument—if you let me phrase the question." Here Pittston had chosen the safest battleground by phrasing the question in procedural terms.

We couldn't tell Judge Christie he had to pierce the corporate veil to save us from a Logan County judge and jury. Lawyers don't speak that way. But the disaster had received widespread publicity, as had the filing of our case. And Judge Christie was an old Democratic politician, the former political leader of Mingo County. President Kennedy rewarded him with this federal district judgeship for bringing in the votes in Mingo County when Kennedy defeated Humphrey in West Virginia's 1960 Democratic primary. Mingo County is next door to Logan County and not much different from it. So we assumed he'd know why we wanted to stay in federal court. Our job was to give him some facts to support a decision in our favor.

We told him the Pittston documents showed that Buffalo Mining Company was operated as a part of Pittston—as a "Division"—and not as a separate corporate subsidiary. The sign in front of the coal-washing plant notified the public "Buffalo Mining Company—Lorado Coal—A Division of The Pittston Company." Similarly the heading on Buffalo Mining Company's stationery read "A Division of The Pittston Company." Doesn't sound like much. But "Division" is a magic word. By calling it a division, Pittston had encouraged the public to believe this was now a Pittston operation, and no longer a separate Buffalo Mining Company.

We quoted Mr. Reineke's statement in *The New York Times*: "The responsibility is Pittston's in the long range." Pittston's president had said almost the same thing when he told the governor's Ad Hoc Commission:

> [Pittston is responsible] . . . if you are talking about the responsibility for having an impoundment, or the responsibility that it was on our property, or the responsibility that we owned it, I must say "yes." We were responsible for that.
>
> But if you are speaking of any responsibility for negligence, then otherwise I would have to say "no."

Pittston's president even testified, in answer to a question from another commissioner, that Steve Dasovich, the allegedly independent vice-president of the Buffalo Mining Company, was acting as Pittston's "agent." If he was, we didn't have to pierce Buffalo Mining Company's corporate veil. We could sue Pittston directly for the acts of its admitted agent.

We also argued that Pittston had failed to follow normal corporate formalities in the holding of Buffalo Mining Company's shareholders' meetings and directors' meetings. We were trying to fit our case to Judge Christie's findings in his earlier corporate-veil case. In that case, the corporation had held no board meetings and no shareholders' meetings, and Judge Christie found that to be one of the most significant reasons for ignoring the corporation's plea that it was separately run. We didn't have as good a case on the facts. There were minutes for each annual Buffalo Mining Company shareholders' meeting and for at least seven board meetings during the two years Pittston had owned all the Buffalo Mining Company stock. There appeared to be some minor, technical irregularities in these minutes, but they might not be significant enough to convince Judge Christie that Pittston had ignored normal corporate formalities.

We also argued that it would be unjust to require the plaintiffs to sue the Buffalo Mining Company, since it didn't have $52 million to pay us. Financial inability to pay is sometimes a factor permitting a piercing of the corporate veil, especially if the shareholders have not given their corporation enough initial capital to conduct its business. Pittston had expected this attack. It informed the court, in its motion, that Buffalo Mining Company had a net worth of nearly $7 million, and at least $15 million in insurance. Although that wasn't enough to pay our claims, Pittston asserted that our claims were "grossly overexaggerated."

True to our legal training, we too sought refuge in cold, unemotional procedural arguments. We said the court could not, or at least should not, rule on Pittston's motion until we had a chance to depose Pittston's people and learn all the facts. Under the Federal Rules of Civil Procedure, a court can rule on a preliminary legal question if the facts bearing on that question are admitted or undisputed. But we claimed that the material facts of this corporate-veil question were still in dispute. Indeed, we wouldn't know what the facts were until we deposed Pittston's people and learned firsthand how the Buffalo Mining Company really was operated. This was our fall-back position. If all else was lost, at least this argument could keep us in court awhile longer.

Within a few weeks after the filing of our response to Pittston's motion to dismiss the case, Mr. Staker called to tell me Judge Christie had decided to remove himself from the case. Apparently Judge Christie felt compelled to do this because the president of the Pittston Company was one of his oldest friends. I was upset that this news reached me not from the judge, and not through our local counsel, but from my opposing counsel. This made me feel even more an outsider in a coal-company land. But more important, I was embarrassed that all our research hadn't turned up this close friendship between Judge Christie and the president of Pittston.

I was depressed. Our strategy was in shambles. We had made a major mistake. We had gone out of our way to pick a judge who then turned out to be a good friend of the Pittston Company's president. The survivors sure didn't need to go all the way to Washington for that kind of lawyering.

Judge Christie, having removed himself from the case, sent the case to the next most senior judge in the Southern District of West Virginia, Judge Knapp, a recent Nixon appointee. We immediately tried to find out as much as we could about him, but since he was only recently appointed, he had written few opinions. We did learn from local lawyers that he did not move cases along rapidly. This would create a problem for us, since we could not afford a judge who would permit Pittston to delay and drag out these proceedings until our people had to cave in.

To our surprise, Judge Knapp decided that he was too busy to work on our case. I had thought a West Virginia federal judge would be eager to handle a case involving so many sympathetic victims. Certainly the case was more interesting than the normal federal court cases these judges sit through—interstate car theft, land condemnation, and so on.

Now the case was put in the hands of Judge K. K. Hall, the only other federal district judge in the Southern District of West Virginia. Since there was no other judge in that district with less seniority than Judge Hall, he had no choice but to accept the case. But that did not mean he would have to keep the case in the federal court. He could now read the formal, legal papers and dismiss the case, sending it back to a state court in West Virginia, if he wanted to.

We scurried around trying to find out about Judge Hall. Again we had to search for legal opinions. However, as a recent judicial appointee, he had written only a few opinions as a federal judge. We read all of them but gained no real insight into him. Local lawyers were more helpful. They told us Judge Hall was fifty-five years old, a lifelong Democrat, and one of the closest friends of West Virginia's Senator Robert Byrd. This friendship may explain why President Nixon, a Republican, had appointed Judge Hall, a Democrat, to this judicial position. Apparently, this was one of a number of moves President Nixon made in an effort to win the support of Senator Byrd, the Democratic majority whip in the Senate.

Judge Hall was well qualified to be a judge. He had been a Social Security hearing examiner for a few years, and before that a state court judge for sixteen years. He had risen from poverty to judgeship, and had an understanding of the little people's problems. He was solid, a "let's get the work done" kind of person, a very practical man with great common sense. He quickly accepted his duties and set a hearing date for Pittston's motion to dismiss our case.

Before this hearing, we had another meeting with Pittston's lawyers. Again they brought up settlement. Pittston could try to buy us off cheap now, knowing that we had to worry whether Judge Hall might grant Pittston's motion to dismiss our cases. If we waited, and lost that motion, we'd be left with no negotiating position.

Mr. Staker began: "Buffalo Mining Company is now, as it has always been, willing to sit down and discuss these claims to arrive at a fair and just settlement. I'm disappointed you haven't responded to my earlier offer to discuss settlement."

"You have a right to know how we feel about settlement," I said. "We believe these people are not in this case just to recover damages for their houses and personal property. They brought this lawsuit for a bigger purpose, because they want to change the way in which they live with the Pittston Company and other coal companies.

"There are three major elements in their request for relief. Compensatory damages, punitive damages, and injunctive relief. They hope that recovery of substantial damages and an injunction will cause Pittston to change its conduct."

"Well," Mr. Staker quickly responded, "I assure you Judge Hall will have a marked reluctance to issue any injunctive orders in this case. And of course there is no basis for injunctive relief, since the Bureau of Mines and Congress now will ensure that there are no further disasters." Finally, he added, "I don't regard this case as one for punitive damages, since the conduct involved was not reckless or wanton."

My turn. "As you know, the governor's Ad Hoc Commission has found that Pittston's conduct was in flagrant disregard of the lives and property of the residents of Buffalo Creek. That sounds like reckless disregard to me."

"Now, Mr. Stern, I know all about the governor's Ad Hoc Commission. That commission was constituted to investigate the disaster, but its findings are no judicial fiat and that commission's report will have no bearing on the litigation before this court."

"Well, Mr. Staker, I can hardly advise our clients to discuss settlement with Pittston in terms which will not include punitive damages for reckless disregard, now that the Ad Hoc Commission has found Pittston's conduct to have been reckless."

We were at loggerheads, so we took a coffee break. During this break, Mr. Staker tried to lessen the tensions of our meeting. He said, "I saw *The Ten Commandments* on television last night. I saw that movie many years ago, but I had forgotten how fantastic it is. Cecil B. DeMille is a genius. His movie is so vivid and so authentic that you forget you are watching a picture. He has one scene, when the Jews are crossing the Red Sea, with the waters parted momentarily, that is phenomenal. He shows a wall of water lapping at these people from each side, with roaring waves and the water rolling and threatening to break through at any time. I could hardly believe it was not real."

I listened to this story with growing anger. It was obvious that Mr. Staker had no understanding of how involved I had now become with the suffering of the Buffalo Creek people. I was appalled that he could tell me this story without realizing, in any way, its parallel to the Buffalo Creek case. I was determined to hurt him for it. When he finished his Red Sea tale, I quickly responded, "That really was an act of God, wasn't it." His face showed I struck home. I hoped he understood I was not about to be moved by his pleasantries.

Lawyers are not supposed to become this emotional about their cases, but this was not a typical one. This case was a serious matter for me, and as I got more into it, it became even more serious. I was beginning to identify directly with the suffering of these people. I suppose I suffer as a "survivor" myself, as a Jew who survived the Nazi concentration camps, although I wasn't in a camp. In fact, I was only a small child in America at that time, and I don't know of any relatives who died in the concentration camps. But as a Jew, I still feel lucky, and guilty, that I survived while other Jewish children my age died.

I also identified strongly with our plaintiffs' rage at the indifference of the coal companies and their lawyers. I get upset when people ignore me or others. This may be one reason why I joined the civil rights battle, to help the voiceless blacks. I know I always root for the underdog. I never liked the New York Yankees when I was a kid because they never seemed to lose.

Anyway, I was determined that Pittston and Zane Grey Staker would recognize us before the case was over.

We then returned to our formal discussion. Since it was clear that Mr. Staker did not seem to understand the mental injuries of those who survived this disaster, I decided to point out how important this suffering was if we were to discuss settlement even of the compensatory damages of the case. By this time I had met Dr. Lifton and was beginning to understand the emotional problems which are typical for survivors of disasters. However, Dr. Lifton and our other psychiatric experts had not yet gone to the Valley to see any of the plaintiffs.

I told Mr. Staker, "We have retained a number of experts to study the mental suffering of the plaintiffs. They believe that each plaintiff is suffering survival syndrome. For us, this now appears to be the most significant element of the compensatory damages. We are not in this suit to fight over the cost of nails and broken-down televisions that may have been lost. I assume we will have no problem in agreeing on a fair amount for those items. But that is insignificant compared to survival syndrome, which is what the compensatory damages are all about in this case."

Mr. Staker was ready for me. "Now that we are talking about matters in the area of psychiatry, I want to point out that that is something that people always claim. I do not believe that you can prove any psychotic reactions for which you can collect in this case. The defect in the plaintiffs' position is your inability to show the mental suffering which the law will pay for. The proof of mental suffering requires more than mere puff and blow."

That was the end of the meeting. Although Mr. Staker specifically told me again as we walked out of the room that he was available to discuss settlement with the plaintiffs at any time, I realized we were too far apart even to begin such discussions. We were becoming more and more convinced that the mental suffering of the plaintiffs was probably the most significant element of damages. Pittston viewed them as "mere puff and blow." Similarly, we fully expected to prove that its conduct was reckless. Pittston insisted there was no possibility of any such proof.

During the meeting I had told Mr. Staker I thought their motion to dismiss our case, before we could conduct any significant discovery, stood little chance. Mr. Staker said he was of "a distinctly different persuasion with respect to the frivolity of Pittston's motion." When I told him I would immediately begin the depositions of Pittston's officers, if Pittston's motion was denied, Mr. Staker responded that, in the event Pittston's motion was denied, "we will all have to take another look at the case." So it appeared to me that Pittston believed it was going to win on its motion to dismiss our case. Thus, it felt no pressure to make any kind of reasonable settlement offer.

IX

"Oyez, oyez, oyez"

We now had less than a week to get ready for the oral argument before Judge Hall on Pittston's motion to dismiss our case. Pittston had a surprise for us. Three days before the date set for the oral argument, Pittston filed a lengthy memorandum with Judge Hall setting forth new facts and arguments to support its motion to dismiss—leaving enough time for Judge Hall to read it, but not enough time for us to respond in writing.

This new memo and affidavits were devastating. Pittston attacked our "super-technical quibbling over whether Buffalo Mining Company's attention to corporate formalities has been sufficiently punctilious—hardly a basis for overturning the constitutional doctrine of limited shareholder liability." Pittston concluded—relying on new affidavits swearing to the authenticity of the minutes of various Buffalo Mining Company shareholders' meetings—"There is thus not the slightest formalistic irregularity with regard to any aspect of Buffalo Mining's shareholders' meetings."

Division? "Buffalo Mining Company has been inaccurately referred to as a 'division' instead of a 'subsidiary.' " But that was a simple mistake, merely "Buffalo Mining's boastful stationery."

What about Mr. Camicia's damaging testimony that Mr. Dasovich was a Pittston agent? Mr. Camicia had made that surprising admission, with Mr. Staker sitting at his elbow, in answer to a question from Ad Hoc Commissioner Bill Davies. This was the same Mr. Davies who had studied refuse piles in this country for the U.S. Geological Survey after the Aberfan disaster. Governor Moore had made him a commissioner because of his engineering expertise, but he proved to be a pretty good interrogator, too. When I first read how he got Mr. Camicia to admit that Mr. Dasovich was a Pittston agent, I was astounded. I called to congratulate him and asked how he learned the art of cross-examination. He said he'd checked some books out of the library and followed their suggestions. He had all his questions written down to lead Mr. Camicia, carefully and skillfully, to give the answer he wanted:

> Q: Mr. Camicia, you have said that Buffalo Mining Company operated independently. Were they given a distinct assignment or delegation of authority to operate that way with Pittston?
>
> A: Certainly. Yes, sir.
>
> Q: Therefore, Mr. Dasovich was the managing agent, as far as Pittston was concerned?
>
> A: Yes. He was. He has the direct responsibility for that division.
>
> Q: And he was acting as the Pittston agent in this case?
>
> A: Yes.

Mr. Davies also got the same admission out of Mr. Spotte, the president of the Pittston Coal Group. Pittston had to find a way to explain these answers. It did. Its memo to Judge Hall said:

> [T]hese brief answers, given by nonlawyers in response to a question formulated by a geologist on a subject not further pursued by him, are irrelevant on the existence of agency in the technical, legal sense.

Up to this point I had been preparing our written responses to Pittston's written arguments. But now we were nearing the time when the lawyers would appear before Judge Hall to argue orally the merits of their positions. We had to decide who would make the oral argument for our side against Zane Grey Staker.

Mr. Staker has a beautiful, sonorous voice, a magic way with words, always using two syllables if one will do, or ten words if five will do. I was told he'd never lost a jury verdict. He is pure theater, and the folks on the county juries love him. He's extremely polite, very well mannered, a Victorian at heart. He comes from a fine family, and his father still lives with him on the massive spread where Zane raises cattle. But Zane also is a former naval commander, with all the habits he learned there—smokes too much; drinks his coffee black, and often; wears the same uniform every day, gray suit and black tie; and likes to command attention. He's used to having his way. And he's smart enough to get it. He's a Harvard Law School graduate and a former president of the West Virginia Bar Association. Lots of coal-company clients. And that name. Zane Grey Staker. He doesn't like to be called Mr. Staker. He always thinks his dad is in the room when he hears someone say "Mr. Staker." No. It's "Zane Grey Staker." It rolls out of the voice box like thunder and strikes terror in the hearts of the opposition. Who would argue against him, this distinguished-looking, gray-haired, fifty-year-old West Virginia legend?

Not me. I was only thirty-five years old. Too young-looking. Even had a beard. And I was an outsider. But I could prepare Bud Shay to argue for us. Bud is older, wiser, and was first in his class at West Virginia University Law School. He has all the right credentials, and no beard. Indeed, he has a crew cut. But Judge Hall set the hearing at a time when Bud and his wife would be in Bermuda on a long-promised vacation. Bud's wife would be heartbroken if they had to cancel the trip. So it looked as if I was it. I shaved my beard.

Luckily, though, I then found a way to get Bud and his wife to Bermuda without missing the hearing. So Bud was it again.

It was now six months since the complaint had been filed, yet this was the first time any federal judge had held a hearing or even met with the lawyers in the case. Pittston's lawyers had done a good job, to date, in delaying matters.

The judge's door opened, and we all jumped to our feet as a kindly-looking, gray-haired man in black robes, the Southern District's most junior judge, strode swiftly into the small federal courtroom. We remained standing, facing Judge Hall, as the clerk announced:

> Oyez, oyez, oyez. Silence is now commanded under penalty of fine and imprisonment while the Honorable K. K. Hall, Judge of the United States District Court for the Southern District of West Virginia, is sitting. All persons having motions to make, pleas to enter or actions to prosecute or defend will now come forward and they may be heard in their turn.
>
> God save the United States of America and this honorable Court.
>
> Please be seated and come to order.

Judge Hall had carefully designed this unique opening to ensure that his name would not be mentioned twice. A usual beginning would have the clerk announce, "Present in the Courtroom is the Honorable K. K. Hall" before beginning with the "Oyez, oyez, oyez" and the repetition of the judge's name. In West Virginia's state courts, where Judge Hall had been a trial judge, the judge's name is mentioned only once, when the clerk announces the opening of court each day. Judge Hall preferred it that way. His judicial robes did not hide the man underneath.

The hearing got under way with a few preliminaries. Judge Hall wished us all a good morning and asked our local counsel to introduce the out-of-state counsel, as is customary. Bud Shay introduced the Arnold & Porter lawyers to Judge Hall, who smiled down at us and said, "You are welcome to appear before this Court in this case, on this matter, and on all other matters that might arise in connection with this case."

My heart leaped. We'd won. There would be no need to permit us to appear before him on all matters that might arise in the future if he was about to dismiss our case.

The formalities continued while I tried to restrain my excitement over Judge Hall's welcome. Mr. Staker introduced Pittston's New York counsel, and then Judge Hall listened to lengthy arguments from each side on Pittston's motion. As soon as the arguments ended, Judge Hall ruled from the bench that Pittston's motion to dismiss the case would be denied, for the time being, at least until the plaintiffs had been permitted full discovery on all the issues in the case.

For months thereafter Pittston's lawyers argued with us that Judge Hall had not denied their motion on its merits and had only determined that their motion was premature. Procedurally Pittston was right. But substantively we had won. I could not believe that Judge Hall, with his common sense and his desire to move this case along, would permit the parties to engage in discovery—which would take a long time, including depositions of almost 600 plaintiffs and voluminous document discovery—only to dismiss the case once all this discovery had been completed.

Judge Hall then shocked us again. He wanted the parties to know how he thought the case could be tried. He envisioned one jury, with numerous alternates, who would hear four or five sample cases on the question of liability. This jury, assuming it determined that Pittston was liable, would then be dismissed for a short period of time and brought back later to determine the amount of damages for each of these four or five sample cases. Pittston's lawyers were stunned that Judge Hall had thought all the way through to the trial of the case. Judge Hall even ordered a timetable for the completion of discovery and other pretrial matters which could have us prepared for that trial within eight months.

I can still feel our exhilaration as we left the federal building that morning. We had won a complete and total victory. We had not been dismissed from federal court. And we had a judge who was willing to listen to the sympathetic concerns of our plaintiffs.

A plaintiff's lawyer often is only as good as the judge he gets. If the judge is too busy or too lazy or unsympathetic, your case can languish for months and even years. We once represented an industrial company in its efforts to take over a corporation based in another city. The takeover candidate wanted nothing to do with this forced marriage and went to court to thwart our outsider's proposal. The federal judge sprang to the defense of the locally based, reluctant bride. But the judge did not rule against us. If he had, we could have appealed the case. Instead he just let the hearings drag on interminably, never ruling for us, but never ruling against us. He'd just put his sunglasses on, to protect his eyes from the glare of the fluorescent lights in his courtroom, lean back in his big leather chair, and appear to doze off while we fought in the pits below. He kept us there, day in and week out, until our client finally wearied of his suit.

But Judge Hall was different. He wanted to move our case along. I thought it was now only a matter of time before we would pick up all the rest of the pieces in this chess game. I immediately wrote our clients a "Very Good News Letter." The West Virginia papers also announced the victory—"Judge Rules Pittston Stays in Dam Suit."

PART TWO

X

"A More Definite Statement"

With Judge Hall's denial of its motion, Pittston agreed to begin producing documents that went beyond the corporate-veil issue. Mr. Staker called these the documents on the "meritorious aspects" of the case. A nice phrase. We hoped Pittston's files, and Buffalo Mining Company's files, would prove the merits of our charge that Pittston was responsible for this disaster, and that Pittston's conduct was reckless, not just careless.

I also continued my efforts to get access to the exhibits and documents collected by the governor's Ad Hoc Commission. I'd tried in vain for six months to see that material, but the governor kept it locked in a room, open to no one. Once, while waiting in the governor's huge outer waiting room, I couldn't help noticing the portraits of West Virginia's two most recent governors hanging on the wall. One was now in jail for bribing the jury which tried him for corruption in office. But his picture hung there still, a symbol, to me, of the corruption inflicted on West Virginia by its absentee coal companies. I imagined that some companies didn't mind paying some of West Virginia's politicians to let them drag away most of the state's wealth from under its ground. The owners in New York and the politicians in West Virginia grew wealthy on the state's vast resources, while the people of West Virginia became the poor hillbillies and broken coal miners of American folklore.

This particular morning a group of schoolchildren was on a tour of the State Capitol. They were ushered in quietly to this magnificient waiting room, with its civic lesson hanging in plain view. But the kids' lesson was different. "The carpet on the floor is 27 feet wide, 72 feet long. It was 2 inches thick at the time it was installed. It was made by Mohawk Mills in New York and is the world's largest seamless carpet. Now come along to the Rotunda, and we'll see the biggest chandelier in the country. It is rock crystal from Czechoslovakia, with 6,500 hand-cut rock crystal pieces. It weighs 2 tons and is 8 feet in diameter. It holds 100 light bulbs, each of 100 watts." Look up, look down—but don't look sideways. You might see the walls and really learn something.

After the schoolchildren left I was told again that I would not be allowed to see the documents and exhibits collected by the governor's Ad Hoc Commission. Another airplane flight had been in vain.

So, at the end of that first day's hearing before Judge Hall, knowing that the Charleston press was in the courtroom taking notes, I asked Judge Hall to help me see the documents and exhibits collected by the commission. He responded, "Well, the Court feels that you should have access to them . . . the Court will lend its power to doing whatever is necessary if you can't work it out." That was all I needed. The federal court didn't have to issue a subpoena to the governor. I was ushered into the Ad Hoc Commission's locked document room that afternoon.

I'd been having the same kind of problem with Pittston, trying to get copies of all the documents on Pittston's insurance coverage. Pittston had said it had about $15 million in insurance which would cover damages from the Buffalo Creek disaster. But the insurance carriers had reserved the right to assert, after all the damage payments were made, that Pittston's insurance contracts did not cover this disaster. If the insurance carriers then prevailed, Pittston would have to refund whatever money the carriers had advanced them. I wanted a copy of this postdisaster reservation-of-rights agreement. Pittston refused. So I asked Judge Hall for help.

> The Court: What would be objectionable about it, Mr. Staker?
>
> Mr. Staker: The fact of certain understandings that have been concluded between the Pittston Company and the insurers as to the possibility of events subsequent to the time that the coverage is paid out. That has nothing to do with the availability of the insurance coverage itself.
>
> The Court: Well, as I understand it, you will be willing to send a copy of that to the Court and if the Court feels it is proper it could forward or send a copy to plaintiffs' counsel?
>
> Mr. Staker: Yes, your Honor.

Soon thereafter Mr. Staker decided not to bother the court with this problem. He mailed the agreement on to us.

The preamble to the agreement was to prove helpful later on. It said:

> [T]here are certain differences between The Pittston Company [and the insurance companies] relating to the negotiations for the placement of coverage and the coverage afforded by certain policies of insurance . . . as well as a policy of insurance placed on February 24, 1972. . . .

What were these "differences" with respect to the "negotiations for the placement of coverage"? And why was one of the insurance policies "placed on February 24, 1972," only two days before the disaster? It would be some time before we found out.

Judge Hall's schedule provided that depositions would be taken between May 1 and September 1, so we wrote to Pittston's counsel that we wanted to depose Pittston's president, Mr. Camicia, and vice-president, Mr. Kebblish, along with seven other Pittston or Buffalo Mining Company employees, beginning on May 1, preferably in Washington, D.C., at our law firm's office.

We also completed preparation of a more definite statement of our damage claims, in response to Pittston's motion to require that our complaint more particularly state the damages suffered by each of the plaintiffs.

We had been hard at work on this task for many months, collecting information on each plaintiff's losses. We had decided to use this information, especially each person's recollection of the day of the disaster, to turn this more definite statement into a compelling story of mental and emotional suffering, to begin providing the details of psychic impairment and survival syndrome.

We filed with the court, and served on Pittston, the stories each disaster victim recalled. They filled two volumes the size of telephone books. It was all there. We left nothing out. We could never have filed such a lengthy document as our original complaint. But we could file it in response to Pittston's insistence that we more specifically tell them what the case was all about. It was not a typical legal document. It spoke to the heart, not the brain. It told the story of this disaster from the personal point of view of the plaintiffs who suffered in it. It was a devastating document, as Dennis Prince's story shows.

I had heard of Dennis Prince long before I met him. His ordeal was known by many in the Valley. At one point, Charlie Cowan pointed him out to me as the man who, after the disaster, often came into the gas station, bought cigars, and then just stood around staring into space. He would light up a cigar and look over toward me, but never speak. After a number of trips to the Valley, I told him I would be willing to talk with him, at any time, once he felt like talking. He said he just couldn't talk about the disaster yet. One day he was ready. I tape-recorded his story.

"The thing came with no warning.

"They said they warned the people, but there was no warning at our camps. The only thing that I saw as I looked at the creek was black set in. Then my wife started hollering, 'Hey, there's a housetop going down the creek.' I ran to the door and saw a bunch of trash in the creek. The housetop had pushed trash in front of it.

"The water was in the road, running down the road across the ballfield. So I told the kids to get out of the house and take nothing with them. I said, 'The dam's broke and the hills are the only chance we have to live. Get out of here. If we don't make it to the hill we'll all be drowned.'

"I got the two young ones, eighteen-year-old Tommy and thirteen-year-old Dawna, and headed them toward the hill. I told my wife that we had to make it to the hill—that that was our only chance. We got out in the yard after the children. The water was running through the ballfield, but I thought we could wade the ballfield and get across to the hill.

"My wife and I had no more got ten feet away from our trailer than the water just raised up, took the underpinning out from under the trailer and wrapped around our legs, and about knocked our feet out from under us. We were in water waist-deep.

"At that point my oldest son, Dennis, Jr., who never did get out of the trailer, yelled to me and my wife to come back to the trailer when he saw the water coming down on us. But the water was so swift that I couldn't hold my wife, and she began to float out away from me. I yelled to my son, 'I can't get your mother back.' Then I looked back, and I saw the whole mountain, like a mountain of water, coming, houses on top of houses. I yelled again, 'I can't get your mother.'

"Just then my telephone pole, the one I had my switchbox on, fell where I had been standing. It would have hit me right dead center on top of the head, and I didn't even see it. My wife hollered, 'The wire lines are falling.' At that time I caught up with my wife again, with the water waist-deep, and got hold of her. Then a car came floating out of our garage, and I tried to get her up on the hood of that car. But I couldn't. There was just nothing to hold onto. She would slide back every time.

"Then I saw Mr. Banks's house come out through the field down below the garage and change its course and start coming across the field. Mr. Banks was my neighbor, and his six-room house hit that car which I had been trying to get my wife on. That knocked the car loose from us. When the house hit the car it just sank it. If we had been on the car that would have been it right there.

"I then jerked my wife back out of the way of the porch of the house as it went by and told her, 'Get on the porch. Get on here.' But just before I told her that she said, 'I know my leg is broke.' I think she broke her leg when the car hit her or the porch hit her. I said, 'Honey, you got to help me with the one you got.' So then I put my arm around her, but the house was traveling so fast out through there that we missed the front. I told her to get the timing right so we could jump and catch the last part of the house. We just managed to grab onto the last post on the house. It hadn't been put on there tight, but it held enough to hold us.

"Then I told her the only hope we got, 'We got to get on top of the house.' So I tried twice to put her on the porch, but she never did have too good a grip with her hands and she couldn't hold without getting a second lift. I'd get her up there and then when I'd turn to get another hold, she'd slide right back down into the water. The water was crawling right on up to her body, and I knew I had to do something.

"You weren't allowed a minute for nothing. Everything was down into the seconds. There was so much noise from this thing you had to yell to talk. It sounded like you were talking into a continuous thunderstorm, ripping up the camp, just tearing these houses all to pieces like thunder. So we were yelling at each other.

"I then yelled to her again, 'This is a must. I got to get you up there this time.' I mean the whole camp was coming at us. I had no time to look at the hill. I didn't know what had happened. I just knew we were traveling at a speed down through there. So I got off the house to help my wife. I went into the water under her and climbed up the pole with her. I knew I had to get her up there on the house if she was to live. I knew that. But when I hit the water it just stretched me out. Had I turned loose of the little metal pole I was holding onto, I'd never catch the house again. So then I pulled myself back through the water and got on the porch.

"I did not have time to explain anything, so I put her hand high on this metal pole and I said, 'Honey, hold on, I'll get you in just a minute. I've got to go first. Don't you turn loose for nothing.' I put her hand there so I knew where it would be. I figured if the water would come over the top of her head I'd still know where she was by where her hand was.

"So then I leaped up on top of the roof of the porch and turned around to get her hand. But in the few seconds it took just to get up on the porch roof something hit her and knocked her back against the wall of the porch, and I couldn't reach her hand. It was only eight to ten seconds at the most that it took me to get up on the roof and how she got back against that wall that fast I don't know.

"At that time she just smiled and waved good-bye to me.

"I ran over to the edge of the porch and tried to get her attention, but I never could. When I laid down to reach and get her, something hit me on the foot. This garage that I had rented had a wide carport on it, and it had come over that roof and was raking me off. So I moved away a bit.

"As soon as the garage moved off just a little, say just ten inches away from the roof, something gave me room to get in there, so I moved back in there to try and get her attention again. Just as I was going on down to reach my arm down in there to tell her to give me her hand, I was hollering at her, the tin garage collapsed and I saw her hair. She was forty-two years old. Her hair used to be red, but it had turned and she looked more like a blonde.

"I saw her hair come up through the black water that was out there and I kept running. I knew she'd come up directly, and I'd get her. Then maybe she went under, and I kept watching for her hair to come up. But she never did."

At that point the house Dennis was on began to pick up speed after a big wave of water came up over the roof. Dennis managed to get up to the top of the piece of the house still above water. But then the house hit a cliff.

"It just went *shhh*, just sheared the porch off just like a meat slicer. I didn't see a board or splinter fly. The house then went on down traveling at thirty-five or forty miles an hour, with me stuck on the rooftop."

Then the house headed toward some trees, "and just folded up. I looked back when the chimney went, and the flying bricks looked like they were about ten feet high in a cloud of cement dust where the bricks were plastered together.

"There I was. The porch was gone and half of the house was gone. I looked down and saw water in the rafters. I then slid down the house and put my hand where the chimney had been. There was a gap in the house there, but I didn't put my fingers down too far, since I was afraid a board or something would get them."

The house then came to a bridge. "There was a wall of boards and lumber jutting up there, and the water was just beginning to spill over each side of the bridge. It was a long way to the hill on one side and a long way to the hill on the other side from where I was. Because of the speed I was traveling I felt 'Well, this is it. I'll just jump to the highest fence post sticking up there. I'll get down like I was about to make a tackle.' Then the current picked up the house and the bridge sheared the front end of the house off.

"The water then floated the house on top of a railroad trestle and I hit on the other side. There was plenty of water on the other side and it wasn't too bad a ride. At that point the house and water had reached the camp where my oldest daughter lived. But the noise was so loud it was useless to holler because nobody could hear me anyhow."

Another bridge then sheared off another part of the front of the house. "I fell for ten feet because there wasn't too much water on the other side of this bridge. The housetop came up after it fell, and I was down in the hole looking up at the daylight and all this black water. There wasn't that much water, so as I was getting up the water collapsed the housetop on me. Each end of the housetop was sticking up in the air. I was afraid it would break any minute with me down in the bottom. As I got back up to the top of the housetop, standing up on it, the bridge broke just behind me. This released a lot of water and I felt the house pick up speed just as I looked back. I fell down in the house again. I had to go down fast to stick on the house, what was left of it. Then I saw the bridge with tumbling automobiles sticking up out of the air with the bridge just warping and everything and I was just in front outrunning the bridge and all this debris. It scared the devil out of me.

"I then looked down and saw the next bridge coming up about thirty feet high and it looked like I was heading straight into it. At this point I was right in the middle of the creek and I began to get very scared. I ain't a frail man but this really scared me. At that point the trash in the water throwed me over into the curb next to the railroad track which had been on the left of the creek as you go downstream. I was missing the main current which was heading directly into that bridge down there. Then something came across that house and jogged me back into the stream. I was getting ready to jump, but this knocked me off balance and I didn't go. Just as soon as the house straightened up and came back around a little more, I took another run and this time I was praying. When I turned around I must have jumped what seemed like twenty feet, clearing the water up onto the grass by about eighteen inches. I didn't have my shoes tied and I lost my right shoe. I turned around to put my one shoe on and looked to see what had happened to the housetop I had been on. I saw the housetop just disintegrate into the bridge down about a couple of hundred feet below where I was."

Dennis then went searching for his children. He found Tommy and Dawna up on the hill back where they lived. Tommy had managed to get Dawna up on the bank just before the corner of a house came through and hit him in the back. Tommy then made it to the bank but he was in bad shape. Eventually a helicopter came and took him to a hospital. The doctors had to remove sixty-three inches of Tommy's small intestine.

Although Dennis, Jr., didn't make it out of the family's trailer before the water took it, the trailer only moved a short way down the Valley before it came to a stop by the hillside. He walked out unhurt.

Mr. Prince spent the next nine days hunting for his wife, Margie, digging through trash piles and everything else, hoping to find her alive. He kept hearing about people finding people alive, in automobiles, digging them out of drifts, and so on. He asked everyone if they'd seen her. Everyone knew her. "She was an awful nice person. She was a cash-out lady at the Island Creek store." Nine days later her body was found under some splintered homes and trash at Amherstdale, some five miles down the Valley from their home at Lundale.

I saw Mr. Prince's obvious suffering as symbolic of the

suffering of all the survivors. That is the reason I captioned their Complaint, DENNIS PRINCE, et al., Plaintiffs, vs. THE PITTSTON COMPANY, Defendant, listing Dennis Prince as the first of our hundreds of plaintiffs.

In addition to the individual survivors' stories, our more definite statement quoted, almost verbatim, the findings of our psychiatric experts. By this time Dr. Robert Lifton and two of our other experts, Dr. Joseph Brenno and Dr. Robert J. Coles, had been to the Valley to talk to some of the plaintiffs. Dr. Lifton reported that "the psychological and psychosomatic effects of this disaster resemble those I encountered in Hiroshima. Though a flood experience can hardly be equated with exposure to an atomic bomb, the survivors of both had much in common, which itself is an indication of the extraordinary human destructiveness of the Buffalo Creek disaster."

Dr. Lifton explained that every survivor of the Buffalo Creek disaster had some or all of the following five manifestations of the survivor syndrome.

The first manifestation is *death imprint* and related death anxiety. The death imprint consists of memories and images of the disaster invariably associated with death, dying, and massive destruction. This left them haunted not just by death but by grotesque and unacceptable forms of death. They have a strong residual death anxiety, with a sense of being fearful and extremely vulnerable, and of the uncertainty and unreliability of any form of life. They view their overall environment—including nature itself—as threatening and lethal rather than life-sustaining. To us, the rhythm of raindrops is like music, bringing life to the spring flowers. To them it is the drumbeat of death, bringing fear of another dam failure, of flood, or death.

Death guilt is the second manifestation of the survivor syndrome—the survivors' sense of painful self-condemnation over having lived while others died. The survivors are plagued by the feeling, however irrational, that they could have or should have done something to save close relatives who perished. The survivor also feels that his life was purchased at the cost of another's—that the other person's death permitted him to live. People who have gone through this kind of experience—the soldier who sees his buddy killed, the child who sees cancer strike all around him in his family, the lone survivor in a car crash—are never quite able to forgive themselves for having survived. On the other hand, they can't help but feel relieved that they had the good fortune to survive, which only intensifies their guilt.

Psychic numbing is the third category. It is a diminished capacity for feeling of all kinds, in the form of apathy, withdrawal, depression, and overall constriction in living. Psychic numbing is perhaps the most universal response to disaster. Partly it is an extension of the "stunned" state experienced at the time of disaster, a defense used to protect the mind from the full impact of the death all around. The numbing persists at Buffalo Creek because the people still need to defend themselves against their death anxiety and death guilt. They withdraw from groups, from activities of various kinds, from each other. Or they do just the opposite, engage in overactivity, like Roland Staten going back to work in the mines until physically exhausted, until he could find mental rest.

The fourth category is that of *impaired human relationships*. With so much of their lives so suddenly and totally destroyed, the survivors feel victimized, much in need of help and assistance, but at the same time highly suspicious of any such affection or help because that only reminds them of their own weakness. They have a special need for family closeness, but also a particularly strong tendency to grate on one another in a way that intensifies the others' resentment and fear. Marital breakdown, sexual impotence, anger and rage they cannot express, an overall sense that everything, everyone, is suspect and life itself is counterfeit, are common manifestations.

Finally, the survivors search for a way to give their death encounter *significance.* In many disasters survivors are able to find some comfort, or at least resignation, in the deep conviction that what happened was a matter of God's will or of some larger power that no mortal could influence. But this disaster was man-made, and such survivors can find no reason or justification for it. Their suffering seems unresolvable.

Dr. Lifton felt that no one exposed to the Buffalo Creek disaster escaped the significant psychological suffering associated with these five patterns and conflicts common to such disasters. But the Buffalo Creek disaster was even worse than other disasters. It was unique in its combination of suddenness, destructive power within a limited circumscribed area, and resulting breakdown of community structure. It wasn't a tornado that knocked out three blocks of a city, or an Aberfan that killed the children in the school but left the rest of the community intact. The Buffalo Creek disaster destroyed everything, the entire community. There was nothing left to build on, no roots left to grow again. You couldn't just move to the next street or next town in the Valley. The Valley was gone.

When we filed our more definite statement of the plaintiffs' damage claims, we also amended our complaint to add almost 200 more plaintiffs, bringing the total to 625. Our damage claims, including those for these additional plaintiffs, now totaled $64 million, rather than $52 million.

At the time of the filing of our more definite statement and amended complaint, we got a call from Ben Franklin, a *New York Times* reporter who covers matters relating to Appalachia. He wanted to talk with us about the lawsuit. I was eager to have *The New York Times* write articles about our case. Such articles would not have any influence on a jury in Charleston, since that paper is not read by many, if any, people likely to become jurors in West Virginia. But it is read by many Pittston shareholders, and by Pittston's

New York officers and directors. I wanted these people to know what was happening in this lawsuit. I wanted pressure on Pittston to make fair and reasonable settlements. I was not satisfied that Pittston's New York management knew how horrendous this disaster really was. I realized they did not have any understanding of the survivors' mental suffering if they were getting their information on our "mere puff and blow" claims from their lawyer.

But I was a little wary about talking to the *New York Times* reporter because the lawyers' canons of ethics impose limitations on conversations a lawyer may have with the press while a case is before the court. In general, a lawyer is not supposed to try his case in the press. However, this case, and the incident the case grew out of, had already been the subject of extensive publicity in the press. There really was no way we could keep our public filings with the court from being reported by the press. So we decided to help Ben Franklin, and any other reporter who called, by giving them copies of any documents on public file in the federal court, rather than making them go to Charleston to see those documents. We gave Ben a copy of our complaint, our various motion papers, our newly amended complaint, and the lengthy, more definite statement of the damages suffered by the plaintiffs.

Ben promptly wrote an article for *The New York Times* headlined "Flood Survivors Sue Pittston Co.—Plaintiffs Ask $64 Million—Seek Damages Over 'Survivor Syndrome.' " To the public, this was now a suit about "Survivor Syndrome," about the suffering of the survivors, rather than about the destruction of homes and possessions.

The article had its effect. Pittston's management immediately received phone calls from stockbrokers and shareholders wanting to know what this lawsuit was all about, what survivor syndrome was, and why the suit now was $64 million rather than $52 million.

Pittston was not pleased with our more definite statement,

especially since we still did not specify particular dollar losses plaintiff-by-plaintiff. So they objected to our statement and again asked the court to order us to state our dollar damage claims with more particularity. I felt this was a tactical mistake on Pittston's part. As a purely legal question, Pittston was right; we hadn't specified each plaintiff's dollar losses, item by item. But, to the extent that Pittston really wanted to know this, it had the power under the Federal Rules of Civil Procedure to determine it with scalpel-like efficiency using its own deposition questions and/or written interrogatory questions.

However, by filing objections to our more definite statement, Pittston forced Judge Hall to read our lengthy disaster stories so he could determine whether we had sufficiently alleged each plaintiff's damages. Often judges never get around to reading the legal filings until they are required to do so when parties file motions about them. So Pittston's new motion meant we were able to get our survivors' stories to Judge Hall very early in the case. He now knew we felt that mental suffering, survivor syndrome, psychic impairment, or whatever it was called, was the most important element of our compensatory damage claim.

extent to which human elements
(bored judge) matters

XI

"Pittston Loses Bid for Secrecy"

I had already notified Mr. Staker that I wanted to depose nine Pittston employees. I now wrote him to add the names of twelve more of its employees, and I also told him I wanted the depositions of all twenty-one of these people to begin on May 7 in Charleston.

Pittston responded with a notice that they wanted to depose twenty-one plaintiffs beginning May 7 in Charleston. They also sent us written interrogatories, asking for all kinds of minutiae from each of the 625 plaintiffs, to be answered within thirty days as provided by the Federal Rules of Civil Procedure. A week later, they used another one of these rules. This time they asked that we present each of the plaintiffs for a medical and psychiatric examination by a Pittston doctor in Williamson, Kentucky.

This was a two-pronged attack. First, Pittston would keep us so busy answering their requests that we couldn't do a thorough job in our own deposition efforts. We would have to stretch our manpower to schedule and prepare the plaintiffs for their depositions, to prepare interrogatory answers, and to schedule the plaintiffs for their medical examinations. Second, this would bring pressure on the plaintiffs to get them to settle rather than go through depositions and medical examinations. Up until this point, the plaintiffs had not had to do anything other than respond to our requests for information when we visited the Valley. Now, however, they would have to go to Charleston, over a hundred miles from Logan County, to be questioned under oath by hostile lawyers. In addition, they would have to go over a hundred miles in the other direction to Williamson, Kentucky, undergo a physical examination, take a battery of psychological tests, and be questioned by a psychiatrist.

We had no choice but to object to Pittston's attack. We filed a motion with the court asking that the depositions be held in the Buffalo Creek Valley at the Accoville Grade School rather than in Charleston. This would be more convenient, and less upsetting, for the plaintiffs. We also objected to the requirement that the plaintiffs go all the way to Williamson for medical examinations. We argued that there were facilities for such examinations at the Man Appalachian Regional Hospital at the mouth of the Valley. And it would be a lot easier to send Pittston's doctor there than to send 625 plaintiffs to Williamson.

Judge Hall set a hearing for May 16 on Pittston's motion for a more definite statement and our motion to change the location of the depositions and medical examinations. Two days before this hearing, Mike White, a young, bushy-mustached reporter for the *Charleston Gazette*, wrote a lengthy story on our more definite statement. His story, "Minds Can't Forget Buffalo Creek Dam," appeared on the well-read front page of the Current Affairs section of the *Gazette*'s Sunday paper. It was a lengthy compendium of the survivors' mental suffering, made all the more vivid by pictures from the Valley.

The story began with these words from one of the survivors as they appeared in our more definite statement:

> How do you tell how you feel on paper? It won't be easy. For instance, how do you write how you feel when you are at your brother's house (trailer) on Christmas Eve when all of a sudden your sister-in-law begins to cry and you know it is because she lost her sister and nephew in the flood.
>
> Then you go home and can't sleep because you are so full of tears you feel as though you will burst. You close your eyes and see a row of houses as the flood hit. You lie there until 4 or 5 A.M. with all these things going through your mind. You even cry until sleep brings you relief.

Even if Judge Hall had not read every page of our lengthy, more definite statement, he certainly knew its highlights from Mike White's article in the *Gazette*.

Mike White's continuing articles on our filings in federal court, written in easily understood lay terms for his readers, educated the people of Charleston about our case. This was important because the jury eventually would come from Charleston. Later, when it came time to select the jury, Pittston would discover that it was in hostile territory. But we were a long way from any jury selection.

"Oyez, oyez, oyez." The second hearing before Judge Hall was about to begin. Judge Hall opened by requesting counsel "in addressing themselves to the questions presented to limit any argument to not more than ten minutes on any one point and less if possible." Mr. Staker thought Judge Hall's remarks were addressed to him personally, and he seemed on the defensive all day. "I am not going to ignore the Court's admonition at the outset that there be some limitation on utterances made here in connection with these motions," and "again I am going to hush because the Court has said that it doesn't want too much mouthing about here."

One reason Pittston insisted on a more particular specification of damages was to prove that some of the plaintiffs did not have at least $10,000 worth of damages. In a diversity case such as this, each party must have a claim for damages in excess of $10,000 to stay in federal court. This keeps the federal courts from being cluttered with small and insignificant lawsuits even though there may be diversity of citizenship.

Pittston could accept the fact that, for the purposes of our complaint, we might prove that a husband and wife each suffered more than $10,000 in damages if they lost their home and all their belongings. But Pittston felt each child clearly could not have lost $10,000 worth of personal belongings. However, since we were also claiming that each child, as well as each adult, suffered $50,000 in damages for psychic impairment, Pittston had to demonstrate that these psychic-impairment claims were frivolous, or that the damages for any such claims could never exceed $10,000, before it could get the lawsuits by all the children dismissed.

Judge Hall quickly responded: "[There are] facts sufficient to—everyone I read—support a possible verdict in excess of $10,000—there is some nervous or mental damage and you can't say as a matter of law that the damages in those instances are less than $10,000."

We now had a ruling from the judge that our survivor-syndrome claims were not frivolous, and that those damages, as a matter of law, might exceed $10,000. We had come a long way from "mere puff and blow."

But we still had a long way to go with Judge Hall. He suggested we "give some thought to selecting a few representative cases for trial on the question of liability to apply to other cases in the same categories, like perhaps . . . cases where the party was actually swept in the stream, for example, cases where members of the family were swept from their arms or died in their visual area, and others perhaps where someone was just sitting on the bank and saw all this happen and saw a member of the family drown or associates drown."

Our experts told us that all the survivors were suffering, whether they were in the water or not, whether they saw their family members die or not. Indeed, most of our plaintiffs weren't in the water and didn't see anyone die. Some were not even in the Valley that day. A few were temporarily away, looking for jobs, in hospitals, or even in jail. If Judge Hall persisted in believing that there might be a difference in Pittston's liability for psychic impairment, based on the location of the survivor at the time of the disaster, many of our plaintiffs might be unable to collect for their mental suffering. This problem would haunt us throughout the case.

Our efforts to protect the plaintiffs from trips to Charleston and Williamson were unsuccessful. Judge Hall agreed with Pittston that the plaintiffs could be required to go to Charleston for depositions and to Williamson for medical examinations if Pittston would pay them $20 a day plus travel expenses of $.10 a mile. The money requirement helped alleviate the inconvenience. At least the plaintiffs would be paid some money for their troubles. And this was no small amount of money. For a family of four, driving over to Charleston meant a $20 fee plus $24 for travel for each, or $176 for the family—a total of about $25,000 for all the plaintiffs for their depositions and another $25,000 for all their medical examinations.

We also got into the insurance question again during this hearing. We already had the postdisaster agreement between Pittston and its insurance companies, thanks to Judge Hall. Now we wanted the documents relating to the placing of that insurance, especially the insurance placed only two days before the disaster. Pittston objected: "I don't believe under the rules that they are entitled to all of the documentation which relates to that." Judge Hall responded: "Whether they are entitled to it, what's wrong with their looking at it if it isn't going to cause you a lot of trouble?" So Pittston agreed to give us some more insurance documents. We'd soon see if this was all the documentation we would want on this insurance question.

Finally, I forced Pittston to ask Judge Hall, publicly at this hearing, whether it could insist on keeping our depositions of its employees secret. I was sure Judge Hall would rule that this testimony of Pittston's people should be available to the public and the press. And even if Judge Hall ruled to the contrary, I felt the press would ridicule Pittston's efforts to keep this testimony secret. This was the time of Watergate. The Senate hearings were on television daily. I always felt there was a small parallel between Watergate and Buffalo Creek. While the senators probed to find out what caused that disaster, I was probing to find out what caused this disaster. At any rate, this was no time to be asking a federal court to keep information from the public.

By this time we had already been through one week of our depositions of Pittston's president and vice-president. I knew this deposition testimony was most troubling to Pittston. At the commencement of the deposition of Pittston's president, Mr. Staker agreed that there would be no necessity for Mr. Camicia to read the transcript of his testimony after the reporter transcribed it. A party usually waives his right to read the transcript to verify that the reporter has taken his testimony down accurately and typed it properly. However, after hearing some of Mr. Camicia's answers, Mr. Staker had retracted the waiver of the reading and signing of the transcripts. He then asked me to agree to keep the deposition transcripts "sealed" from the public.

I told Mr. Staker that I would not agree to keep the deposition transcripts sealed, but that I would give him an opportunity to ask Judge Hall for a ruling requiring secrecy. To protect Pittston in the interim, I did agree, temporarily, not to make the transcripts available to the public.

So, at the end of this hearing before Judge Hall, and with the reporter from the *Charleston Gazette* in the courtroom, I asked Mr. Staker if he now intended to ask Judge Hall for an order requiring that the deposition transcripts be sealed from public scrutiny. This forced Mr. Staker to his feet.

> Mr. Staker: Your Honor, simply because I desired to avoid any acrid suggestion that motions out of time were being made, I told counsel that we would undertake to bring the matter up. Here is what it is all about.
>
> * * *
>
> I would not have had this thought come to mind practicing law here in this area were it not for the fact that virtually contemporaneously with the filing of the more definite statements by counsel for the plaintiff a Mr. Ben Franklin of the New York Times, the Washington correspondent for that New York Times, armed with, as he said, what was almost a Sears Roebuck catalog of more definite statement summaries, et cetera, was in touch with our client and various officers of the client even before I had seen this thing and knew anything about it, stating that it had been received as we conclude from counsel for the plaintiffs or somebody acting for them, and wanting to know about this and that. That was the New York Times aspect of it.
>
> Here is a full page Charleston Gazette-Mail treatment of the content of various of these family summaries, notably elevated to the dignity—erroneously, of course—of plaintiffs' depositions.
>
> * * *
>
> I have had no—may I admit in my twenty odd years of practice—experience with this before because I have never seen it happen before and I have never had occasion, therefore, to urge upon the Court the need for protection, and again I am going to hush because the Court has said that it doesn't want too much mouthing about here.
>
> The Court: I didn't put it in those terms.
>
> Mr. Staker: I know, your Honor, but I do want the Court to know that I feel that this is abrasively improper

> and not calculated to do anything at all beneficial to the progress of the trial or the ultimate fairness that is intended to and necessarily must characterize its conduct and disposition.

Judge Hall was ready to rule. He immediately said:

> The Court doesn't intend to issue any order such as is suggested. I'm not suggesting that publicity should be sought or anything like that, but anything we do is open and if the newspapers get hold of it and want to print it, that's their business.
>
> We will try to get fair and impartial jurors who are not influenced in any way when we get to the trial of the case, but I don't intend to attempt to gag the newspapers or the other media in any way.

The afternoon Charleston paper, the *Daily Mail*, headlined its story, "Pittston Loses Bid for Secrecy." The story the next morning was even better—"Buffalo Flood Case News Gag Move Rejected."

XII

Some Lawyer Stories

When Pittston lost its bid to keep the deposition testimony confidential, the depositions took on even greater significance. Anything we discovered could become public knowledge and might be used against Pittston anywhere, at any time. Even in Maine. Pittston's Oil Division had applied to Maine's Department of Environmental Protection for authority to build a $350 million oil refinery and tanker terminal at Eastport. Some environmentalists opposed Pittston's application because an oil spill, in those rough Maine waters, would seriously damage one of the finest fishing grounds along the East Coast. Pittston responded that it would take all the necessary precautions to operate safely. Then the Maine environmentalists heard about the Buffalo Creek disaster and began arguing that Pittston's record in West Virginia proved Pittston could not operate safely in Maine. Pittston responded that it had no responsibility for the Buffalo Creek disaster, that it was the responsibility of the Buffalo Mining Company. The stakes were high. If our depositions uncovered proof of Pittston's responsibility for the Buffalo Creek disaster, it might even lose its bid for a Maine refinery.

We had asked Pittston's lawyers to permit us to take our depositions of the company's employees in Washington so we wouldn't have to transport 15,000 pages of Pittston documents to Charleston. Naturally they refused. They didn't want to make it easy on us. So we got the Holiday Inn in Charleston to provide a specially locked room for our documents. We then boxed them up and shipped them to Charleston. There we stayed for our first round of depositions, cross-examining sixteen Pittston people in thirteen days from 9:00 A.M. to 5:00 P.M., with weekends off to catch up. Later, in New York, we deposed the five remaining Pittston people from our first list of deponents.

Each day we'd meet in a small room at the Holiday Inn. The beds were pushed back. Two long folding tables were brought in, covered with green cloth, and arranged in an L shape. Zane Grey Staker and that day's Pittston witness sat behind one arm on the L, I sat behind the other arm, and the court reporter sat in the no man's land between us. This room was the war zone, and it wasn't easy for Pittston's witnesses.

Cross-examination is our system's method for getting the truth. The thumbscrew, or rack, may be quicker, but results obtained that way are always in doubt. When tortured, a person may admit to anything the questioner asks. But cross-examination is also an ordeal. The witness cannot escape the questions. His lawyer may object to the question, for the record. But the witness must answer, unless his lawyer is willing to stop everything and force the interrogator to run to the judge for a ruling. Lawyers rarely do that, since they know judges like to permit broad latitude in discovery. So the witness has nowhere to hide from the prying, hostile thrust of the cross-examiner, who pushes the witness to admit whatever he wants admitted. It is especially difficult for the witness when 125 people have been killed and the lawyer keeps asking him questions that make it look as if he could have done something to save their lives.

I like to cross-examine, to get into someone else's mind, to find out what he really knows or thinks. I don't like not knowing, and I'm not very trusting. I'm also willing to ask questions, and I don't mind, too much, appearing ignorant, asking the obvious, going over a simple answer again and again. So I've got some of the qualifications of a cross-examiner.

I've also learned from watching other lawyers ask questions. My teacher was John Doar during a number of civil rights trials. John was the chief litigator at the Civil Rights Division in the Justice Department when I was there. Years later he showed his litigator's skills to the country when the House Judiciary Committee selected him to prepare the impeachment case against President Nixon. A perfect choice. John is the most diligent, careful litigator I've ever met. That is the first lesson he taught me. Be prepared. The other lessons, such as how best to use documents, are just as simple.

I was sent to Mississippi after John Hardy, a young black student from Tennessee, was arrested in Walthall County. There had been no blacks registered to vote in Walthall County. Hardy and a few other young college students like him were the first kids to come to Mississippi, in 1961, to encourage some of the frightened black people to speak up for their rights. After some effort, Hardy convinced two elderly black citizens to come with him to the registrar's office to register to vote. The registrar was enraged at this effrontery, and ordered them out of his office. They complied, but as they were leaving, the registrar could not control his anger. He hit Hardy on the back of the head with a gun butt. Had the registrar been a sheriff, Hardy would have been arrested for resisting arrest. That's the charge when the wounds show, and Hardy was visibly bleeding and hurt.

But the registrar was not a sheriff and could not arrest John Hardy. So there could be no resisting-arrest charge. The solution: breach of the peace. Hardy had breached the peace by getting his head in the way of the registrar's gun. He was thrown in jail, and I was sent to the South to determine whether there was some violation of federal law. We hoped to prove that the registrar's action was part of a conspiracy by a number of people in this and other surrounding counties to commit violence on blacks who attempted to register to vote. Then, and only then, could the federal government, under a narrowly drafted federal statute, come to the defense of John Hardy.

I found an old man, Mose McGee, who was in town that day and had seen John Hardy come stumbling out of the registrar's office with his head bleeding. He then saw the sheriff come and arrest Hardy a few blocks away for breach of the peace. Mose McGee lived in a little shack, way back in the hills, at the end of a muddy, deeply rutted dirt road. The county powers never paved the back roads in these rural Mississippi counties, unless they led to white men's property. His muddy dirt road, when the rains came, was more than an inconvenience. It was life or death. One day a neighbor's baby, whom he loved and adored, got sick. No doctor could get up the road to them. And they couldn't drive out to get to the doctor. So he bundled the baby up and walked over the hills, for miles and miles, to get to town. The baby died in his arms before he got there.

Mose McGee plowed his fields behind a mule with the plow lines hitched over his shoulders. He did not have enough money to afford a tractor, or even a horse. I came to see him one day when he was working in the fields like that. It embarrassed him. He did not utter a single word. He just unhitched himself from his plow, went into his shack, cleaned up, and then came out. He said, "It's not right for anyone to be seen as an animal. I want you to see me as a human being."

When he saw what happened to John Hardy, he decided to write it down. He knew what had happened to black men in Walthall County in the past, even the spot where the last black man was lynched. This time he hoped he could help by remembering what he saw. He had no paper in his house to write on. The only paper he could find was the envelope from one of those mass-advertising mailings everyone gets. He found an old stub of a pencil and laboriously printed out on the back and then on the inside of the envelope all his memories of this incident. I quoted verbatim from these notes in preparing a sworn written statement for Mose McGee to sign.

John Doar, however, suggested we would be better off paraphrasing Mose McGee's memories. Then when McGee was being questioned on the witness stand in open court by a hostile Southern lawyer, he could reach in his pocket, pull out his notes, and dramatically demonstrate the accuracy of his recollection. This was my first lesson in the use of documents.

Most people, even some lawyers who have had very little litigation experience, believe the best way to use a document is to show it as soon as possible. John Doar taught me to think about when to use a document. Sometimes it is best to produce the document immediately, for shock purposes or for other reasons. Other times it is better to hold the document back and use it to support a witness's testimony, or to prove a hostile witness is lying.

John Doar had a hostile witness in a voting trial in Hattiesburg, Mississippi. In that trial we were trying to persuade three judges of the United States Court of Appeals for the Fifth Circuit that black citizens of Hattiesburg, Mississippi, had been discriminated against when they went to register to vote, primarily through the use of very difficult questions about the Mississippi Constitution. There was also a general question requiring a statement of the duties and obligations of citizenship. Whatever the answer, the registrar usually found the black applicant unqualified. When I reviewed the application forms, I noticed that a white man and his wife had identical answers to this question requiring a statement of the duties and obligations of citizenship. This usually meant the registrar had helped the white people with this question, or that they had helped each other. There was certainly nothing wrong with such assistance, except that the registrar did not provide black people with that kind of help and certainly did not allow black people to help each other.

I sent FBI agents to question this white man to see if he and his wife had been helped, or had been allowed to help each other. He told the FBI they had no help whatsoever. So I did not use him as a witness. However, to my great surprise, the registrar called him as his witness. Apparently the man had been incensed by the visit from the FBI, so he called the defense lawyers and told them he wanted to testify for the registrar. While he was telling the court that he received no help when he went to register to vote, John turned to me and asked, "What do we have on this man?" I quickly told him what little we had and gave him copies of the application forms of this man and his wife.

I would have immediately confronted his witness with the two identical answers. But John was more patient. He first closed every possible avenue of escape. He led the witness into giving the kinds of answers he was intent on giving. Come hell or high water, this man was going to prove that he and his wife had not helped each other with the most elementary aspects of registration, even though there would have been no reason, other than this man's anger, for denying they received some such help.

John asked if he and his wife knew what the questions on the application form were going to be before they went to the registrar's office. "Of course not," he shot back, even though there was nothing wrong with knowing what was on the application form.

"Well, did you and your wife ever discuss any of the possible questions which might be on the form before you went to the office?"

"Of course not."

"When you got to the office, did you and your wife stand near each other when you filled out the forms?"

"Of course not, we were more than ten feet apart from each other."

"Did you talk to each other at any time during the time when you were filling out the form?"

"Of course not."

"Have you and your wife talked about the answers you gave on the form since you filled out the form?"

"Of course not."

"Did you receive any help while you were filling out your form?"

"No."

"Did your wife receive any help?"

"No."

"Sir, let me now show you the application form which you filled out that day. Would you please read to the judges the answer you gave for the duties and obligations of citizenship."

The witness looked to his lawyers and then turned to the judges. It was too late now to try to explain. He couldn't say, "I learned about the duties and obligations of citizenship as a Boy Scout leader, and my wife helped me with the boys, so it is not surprising we gave identical answers." That wouldn't work now. John had boxed him in. So, with a most pained expression, the witness pleaded with the judges, "Must I read my answer?"

The judges said, "Yes, you must."

He then read the answer in a halting voice, no longer the bold, brave, arrogant witness. After he had read his statement, John applied the finishing touches. He asked him to read his wife's statement of the duties and obligations of citizenship. Again, "Must I?" The judges said he must. In a voice drained of all strength, the bluster all gone, the witness quietly and slowly read his wife's statement, a verbatim statement of his own. That was the end of the questioning. Everyone knew he and his wife had helped each other. No sin or crime in normal circumstances, but in a case where blacks were not permitted to help each other, this unintended admission that white people got such help was further proof of discrimination in this county.

I also learned a little bit about lawyering from Thurman Arnold, the founding partner of Arnold & Porter. He gave a speech at Harvard Law School during my last year there. He told us of his early days as a lawyer, long before he ended up in Washington and Arnold & Porter. He had graduated from Harvard Law School and gone back to Laramie, Wyoming, his home town, to practice law. One day a friend came to see him in his small office.

> Doctor: Thurman, I've got a problem.
> Thurman: What is it?
> Doctor: I just received three ties in the mail, three neckties which are very pretty.
> Thurman: What's the problem?
> Doctor: There was a letter with the ties—they said if I liked the ties I should send $3 to them, and if I didn't like the ties I should just return them.
> Thurman: What's the problem?
> Doctor: I like the ties.
> Thurman: Fine, what's the problem?
> Doctor: I don't want to send $3.
> Thurman: Ooooh. I see. Well, just keep the ties. You don't have to do anything.
> Doctor: You mean I don't have to send $3 to them?
> Thurman: No, don't worry about it.

The doctor left, very relieved. Two weeks later he returned, very upset.

> Doctor: Thurman, I just received a letter from the tie company. They say they want me to send them the $3, since it appears to them I have kept their ties.
> Thurman: Don't worry, Doc, I know just what to do.

Thurman then dictated the following letter:

> Gentlemen: The Doctor did receive the three ties which you sent to him. As you have guessed, he is delighted with them and has kept them. Everyone here in Laramie has complimented the Doctor on his beautiful ties. They appear to be well worth the $3 you mention in your letter.
>
> Please find enclosed a pill. The Doctor has prescribed this pill here in Laramie for many years for all the ailments of the body—lumbago, back-ache, arthritis, neuritis, neuralgia, sinus, what-have-you. It is a superb pill, cured hundreds of people here in Laramie down through the years. The pill is well worth the $5 the Doctor charges for it.
>
> Accordingly, please send us $2 which is the balance you now owe us.

The doctor left, excited and happy. Two weeks later he returned, out of breath and even more agitated than before.

> Doctor: Thurman, I just went down to the post office. They had a registered package for me to pick up. It was from the tie company. They've returned the pill. Their letter says they don't want it. And they want their $3.
> Thurman: Well, now. We do have a problem. I know what we shall do. I'll dictate another letter.

So Thurman sent this letter:

> Gentlemen: We are very distressed to learn you did not want the miraculous pill the Doctor sent to you. It truly is a wonderful pill. Cures all sorts of ailments. But we accept your decision not to use the pill.
>
> This morning the Doctor went to the post office to pick up the package you sent him. Here in Laramie such a trip is considered a house call. The Doctor charges $10 for house calls. Please send us the $7 you now owe us.

That was the last time the doctor heard from the tie company.

Sometimes that is all you do in the law. You just keep pressing ahead until the other side cracks.

It was with a little bit of all this learning that I undertook the cross-examination of Pittston's people. I had placed many of them on the deposition list, willy-nilly, hoping something would come up. I placed others on the list because there were documents I had obtained from Pittston which I intended to use to force them either to tell the truth or to lie in a big way.

XIII

"What Do You Mean, 'Dam'?"

Nicholas Camicia, Pittston's president, was the first witness I deposed. I had a few limited goals. I wanted him to repeat under oath some of the helpful testimony he had given during his appearance before the governor's Ad Hoc Commission. For example, I had him testify again that Steve Dasovich was a Pittston agent.

I also wanted to educate him about the case against Pittston. Mr. Camicia had already testified, in a prepared written statement before the Senate Labor Subcommittee, that the dam which failed at Buffalo Creek was built in accordance with the normal and customary practices in the building of such impoundments in the coal fields. He emphasized that these types of impoundments had been proven safe in the past. But the United States Bureau of Mines had just completed a comprehensive study into the causes of the Buffalo Creek disaster. They reported that the "results of our review of available information and detailed field investigation of the sites of Dams 1, 2 and 3 confirm that all three embankments were built by methods not in conformance with current practices of the civil engineering profession in the design and construction of water retention dams."

I asked Mr. Camicia if he agreed with this conclusion.

"I am not qualified to answer that because I don't know how it was built actually."

The study said that "inadequate conveyance facilities were provided to carry natural runoff safely past the embankments."

"I have no reason to agree or disagree."

How about the report by Garth Fuquay of the U.S. Army Corps of Engineers for the Senate hearings? Fuquay said, in part, " 'The basic concept of Dam 3 was not acceptable from an engineering standpoint.' . . . As president of the Pittston Company, would you ever permit a dam to be built the way Dam 3 was built?"

"Now that the fact is behind us, I certainly would not." That's progress.

Garth Fuquay had also told the senators investigating the Buffalo Creek disaster that the successful operation of Dam 3 "depended on uncontrolled seepage; unless some happy accident occurred whereby Mother Nature took care of this fundamental error of conception, the dam was doomed to failure from the time the first load of refuse was dumped."

Mr. Camicia did not agree, "because it has been a common practice to build this type of impoundment for water clarification in the coal fields of West Virginia for many years. And I know of no accident, to my knowledge, prior to this occasion."

I pressed Mr. Camicia, "Prior to the Buffalo Creek disaster, did you know there were hazards connected with the creation of coal-refuse piles above ground?"

"I had never heard of any in the area in which we operated . . . Southern West Virginia, Eastern Kentucky, Southwest Virginia."

Had Mr. Camicia ever heard about the Crane Creek Flood in 1924? It was reported in the *Bluefield Daily Telegraph* as a "Tragedy Without Parallel in Mining Annals of Southern West Virginia." The paper said a gigantic refuse pile was pushed down a mountain slope by thousands of gallons of water. Crushed a house, killed seven people in it, choked the creek, buried the railroad tracks under fifteen feet of smoldering ash, rock, and slate.

He said he had never heard of it, even though the incident occurred near his childhood home in West Virginia.

Had he ever heard of the lawsuit that developed as a result of that disaster? It was filed in the federal court in West Virginia, and was even appealed to the United States Court of Appeals for the Fourth Circuit. The court said custom and usage was no defense, the coal company was responsible for those seven lives.

"I've already testified or said I don't know anything about it or never heard of it."

How about the gigantic slate slide in Letcher County, Kentucky, in 1923? Back then they said it was "the largest and most destructive 'slate slide' in the history of mining operations in Kentucky."

He never heard of it. "I was only seven years old then."

Q: What is *Coal Age*?

A: It's a publication for the coal industry.

Q: Did you know on December 9, 1926, in the *Coal Age*, they reported slides and refuse dumps had been the cause of great expense, much property damage, and some loss of life?

A: No. I know nothing of this. . . . I didn't know what *Coal Age* was then.

Well, that *Coal Age* article said, in 1926, "where refuse is dumped in a hollow which is the drain for an appreciable watershed in hilly or mountainous country, trouble can be expected unless some provision is made to take care of the water." *Coal Age* even gave "an example of forethought in this regard" with a picture of a refuse pile in Logan County, West Virginia—of all places. Seems the coal company that owned that refuse pile, the Youngstown Sheet & Tube Company, decided to put a concrete pipe under it "so that there will be no damming of the water nor percolation through the bottom of the pile." Did he agree it was wise, at least as long ago as 1926, in Logan County, to put a concrete pipe under a refuse pile so there would be no damming of the water or percolation through the bottom of the pile?

"Yes," Mr. Camicia admitted.

I was beginning to batter down the custom-and-usage argument. Pittston's refuse-pile dams on Middle Fork had no concrete pipes to avoid damming water. There was no culvert, no spillway, no emergency spillway, no provision to take care of the water. Pittston just dumped its refuse in Middle Fork Hollow, "which is a drain for an appreciable watershed in hilly or mountainous country." As *Coal Age* prophesied in 1926, "trouble can be expected unless some provision is made to take care of the water."

"Were you aware of the fact in 1942, in Buchanan County, Virginia, there was again a major sludge dump explosion which killed a number of people?"

"No, I was not."

There were more examples, but the point was made. Anyone in the coal industry should have known, long before the Buffalo Creek disaster, that refuse piles, especially those which impound water, are extremely hazardous.

My third goal in questioning Mr. Camicia was to let him learn something about us. I assumed his lawyers had been telling him we were young, wild-eyed liberals, out on a lark, but that our law firm soon would rein us in. At thirty-five I was the oldest lawyer at our firm working full time on this case. Our other lawyers on the case were in their twenties and early thirties. This was a unique lawsuit for our firm, and the costs were beginning to mount, already over $50,000 in out-of-pocket expenses and much more in unpaid lawyers' time. But our firm did not hamstring us in any way. We would be spending weeks in Charleston, renting a number of hotel rooms, and paying for court reporters to transcribe the deposition testimony daily. Expenses would begin to increase much faster, and soon would approach over $100,000. Before the case ended, Arnold & Porter paid out almost $500,000 in expenses. But the firm never interfered with us, never forced us to cut corners. Once the firm accepted the case, it was staffed and supported as that of any other important paying client. I wanted Mr. Camicia to get that impression from watching us at work. He soon would see we were sparing no expense merely because the lawsuit entered the firm as a public interest matter.

Mr. Staker was shocked at how quickly we were able to put our fingers on the documents relating to particular witnesses or on testimony given by earlier witnesses which might relate to the person we were deposing. We were using a pseudo-computer system to pinpoint documents on any subject matter by any witness, and we also had computer cards prepared nightly on the testimony of that day's witness. This so nonplused Mr. Staker that he finally had to comment. "I've noticed you are using a computer." I nodded.

We really did not have a computer. We only had a typed card system with seventy-six holes. Each hole was coded—hole 1 might be Mr. Camicia, hole 10 might be a month, hole 20 might be a particular subject. When we pushed a long prong through one set of holes, we could separate the cards relating to a particular witness or subject. Nevertheless, I did not disabuse Mr. Staker of his belief that we were using an actual computer.

A lawsuit is very much like a chess game, as I have already indicated, and both are very much like warfare.

Much of the battle goes to the person who can throw in the most manpower and weaponry. It was my intention to prove in every way possible to Mr. Camicia and Mr. Staker that we could not be worn down by any tactics they might try. In the past, the coal companies had been able to outspend and outlast their opponents, but they weren't going to win this battle that way. They would learn that sooner or later.

During Mr. Camicia's three-day deposition we also learned some things we didn't already know. For example, I asked him about the preamble to the postdisaster agreement between Pittston and its insurance companies. What were the "certain differences" relating to the negotiations for the placement of Pittston's insurance coverage? After much tugging and pulling, Mr. Camicia finally testified that, following the disaster, the insurance carriers had raised "some question about whether the dam on Buffalo Creek was reported." Mr. Camicia said he had responded, "What do you mean 'dam,' that's an impoundment, a water filtration system."

Apparently, when Pittston last renewed its insurance, sometime before the disaster, it had been required to report to its insurance carriers on all its dams. I did not yet know why the insurance carriers wanted that information. But I now knew, from Mr. Camicia's answer, that Pittston had not notified its insurance carriers of the dam at Buffalo Creek when it obtained this insurance. This explained Pittston's problem with its insurance companies, and why Pittston avoided calling its refuse-pile dam a dam. But I did not yet know how this could be of any use in our lawsuit.

We also struck paydirt with respect to the corporate-veil question. Pittston had filed an affidavit with Judge Hall, prior to the oral argument on its motion to dismiss our case. In that affidavit, Pittston had attached copies of the minutes of Buffalo Mining Company's stockholders' meetings and directors' meetings for the two years that it had owned all of

Buffalo Mining Company's stock. These minutes were intended to demonstrate that Buffalo Mining Company was a separately run, independent corporation. That was what Pittston's lawyers told Judge Hall during the oral argument and in their written brief. They emphasized that these meetings of the Buffalo Mining Company's stockholders and directors "had actually taken place."

However, on the morning of the day I was to begin deposing Mr. Camicia, Pittston's lawyers filed a letter with Judge Hall correcting these statements. In this letter Pittston's lawyers admitted that "on Friday last, in the course of conferring with officials of The Pittston Company, it came to our attention that only two meetings, those of June 3, 1970, can be certainly stated to have taken place. On at least some of the other six occasions, no formal meetings were held."

This was a major admission and would be a big help in our efforts to pierce the Buffalo Mining Company's corporate veil. Eventually, we were able to prove that there were no formal, or informal, meetings backing up any of the purported minutes of the six directors' and two shareholders' meetings. With this evidence that none of these meetings "had actually taken place," we could argue convincingly that Pittston had not run the Buffalo Mining Company as a separate independent corporation, and therefore had no right to hide behind Buffalo Mining Company's corporate veil.

XIV

"It Was a Natural Occurrence"

Pittston's number two man was next. John Kebblish had only recently joined Pittston and thus was not very knowledgeable about its operations. However, the logs for Pittston's helicopter—produced pursuant to our document request—proved that he had made a number of visits to the Buffalo Mining Company's operations prior to the disaster. In fact, he was at Buffalo Creek only two days before it occurred. This seemed most unusual. He explained this strange coincidence. Buffalo Mining Company was trying to obtain a permit for an important strip-mining operation, and Mr. Camicia wanted him there to help persuade the state of West Virginia to grant the permit.

Irvin Spotte, the president of the Pittston Coal Group, was also there, as was James Yates, the Pittston Coal Group's chief engineer. This was quite a combination—Pittston's number two man, and the president of the Pittston Coal Group, and Pittston's chief engineer, and Mr. Dasovich, the Pittston Coal Group vice-president in charge of the Buffalo Mining Company operation. There they all were two days before the disaster. Maybe they had talked about the rain and the rising water behind the dam on Middle Fork, while working on the strip-mine permit.

> Stern: During the entire time on the 24th and 25th that you were in contact with Mr. Dasovich did he ever indicate to you at all any concern about the rain which had been falling in or around the Buffalo Creek coal-mining operation?
>
> Kebblish: I wouldn't say expressed concern, but he did mention the fact that it was still raining and the water was rising in one of his ponds.
>
> Q: What else did he say about that?
>
> A: He hoped it would stop raining.
>
> Q: Did he tell you which pond?
>
> A: He and Mr. Spotte were sitting in the front seat and me and Mr. Yates were in the back and he may have mentioned what pond and not being familiar with the operation, I didn't really know which one they were talking about.
>
> Q: He was talking about this with Mr. Spotte in the front seat and you just overheard the conversation?
>
> A: Yes, but I did participate in it. As I remember, he said the water was coming up and—oh, I don't know, it was twelve feet or eight feet or something like that, below the top of the dam. I asked him if he had an emergency overflow in the pond and he said no, and I said, "Well, don't you think you ought to be getting one in?" I don't think he replied, and that was about the extent of the conversation.

This answer was too good to be true. Pittston's top officials were sitting in a car together, at the Buffalo Creek operation, two days before the disaster, talking about the rising water behind the dam and the fact that the dam had no "emergency overflow" system. Pittston's number two man says he told Mr. Dasovich "don't you think you ought to be getting one in." Mr. Dasovich didn't reply. Mr. Spotte said nothing. Mr. Yates said nothing. Mr. Kebblish says nothing more was said. They just drove on to dinner, while the water continued to rise behind the dam.

To this day, however, I do not know if this conversation actually occurred. Mr. Spotte and Mr. Dasovich and Mr. Yates could not remember any such conversation, even though Mr. Kebblish said it took place in the close confines of an automobile. Maybe Mr. Kebblish wanted only to believe the conversation took place. At least it showed some effort on his part to avert the disaster, though not much. Maybe Mr. Spotte, Mr. Dasovich, and Mr. Yates wanted only to forget such a conversation took place. They had ignored Mr. Kebblish's concerns that they get an emergency overflow into the dam. Often in traumatic situations such as this, we can remember only what we want to remember and are truly incapable of remembering what in fact happened —a kind of psychic numbing like that suffered by the survivors. Or maybe it's like Thurman Arnold often said, "The things I remember best never really happened."

This is an example of the impossibility of finding legal "truth." We had sworn testimony that a conversation took place. Did it in fact take place? The answer, the truth, will never really be known. But for our purposes we had a legal truth, an amazing admission of Pittston's knowledge two days before the disaster that there was no emergency overflow to take care of the water then rising behind Dam 3.

The president of the Pittston Coal Group was next. Irvin Spotte was probably the most difficult of the witnesses. To me, he epitomizes what is wrong with the coal industry, which values production above everything else. The daily reports prepared by the lowest-level workers to the top officers at Pittston show the coal tonnage produced per day from each mine, the amount of time the mine is shut down, the amount of time the mine is operating, things like that. These reports do not focus on profits, and certainly not on safety. In short, you get to the top in a coal company by producing, not by saving lives.

Mr. Spotte had worked his way up to become head of the entire Pittston Coal Group. Only Mr. Kebblish and Mr. Camicia were above him in the Pittston organization. He was making $70,000 a year, quite a lot of money for someone living in a small coal-company town like Dante, Virginia, where the headquarters of the Pittston Coal Group were located. He had men working under him throughout Virginia, West Virginia, and Kentucky. He was at the head of his army, and his army's job was to produce.

The head of such an organization, of men working with their hands, often feels he must prove his toughness to keep the respect of his men. Mr. Spotte prided himself on being tough. I saw him arrive from the coal fields to check in at the Holiday Inn the night before his deposition. There he was in his coal-company outfit—khaki fatigues, big boots, looking very much like someone just in from the African bush. He continued his tough-guy attitude while testifying. At the end of his two days of cross-examination, he looked me right in the eye and told me, under oath, that he and the Pittston Company had done nothing wrong in this disaster which killed 125 people. "We did not feel that we did anything wrong. And we still don't feel we did anything wrong. It was a natural occurrence. It was something beyond our control."

I hoped his lawyers would not mellow him by the time of the trial. I felt that a jury might consider his arrogance outrageous. I hoped they would react as angrily as I did to his attitude.

Despite Mr. Spotte's refusal to give an inch in his testimony, he did provide important leads. He kept emphasizing the Pittston line that the dam at the Buffalo Creek coal-mining operation was built in the normal and customary manner in which the coal industry had built such dams for generations. His constant harping on this theme finally forced me to deal with it directly. I hadn't thought of this before, but I should have. I decided to ask Mr. Spotte about the dams at the other Pittston coal operations.

Lawyers often say a cross-examiner shouldn't ask a question he doesn't know the answer to. That is an impossible goal. Even the best-prepared lawyer will find himself in situations where he must gamble and ask a question to which he doesn't know the answer. In those situations, though, the lawyer can still try to direct the witness to answer from among a limited number of possible answers. I call this the Fuller Brush technique.

I was a Fuller Brush salesman one summer when I was still in high school. Fuller Brush men are trained never to ask a question which can be answered "yes" or "no." You don't say to the customer, "Do you want to buy this Fuller Brush?" Instead, you say, "Do you want two or three brushes?" You don't say, "May I deliver these brushes to you next Wednesday?" You say, "Would you like the brushes on Wednesday or Thursday?" Most people respond in the same terms in which the question is asked. We are lazy. It is easier to answer the either/or, two or three brushes, question by saying "two" or "three" than to answer the underlying question and say "I don't want *any* Fuller Brushes. And I don't want them on Wednesday or on Thursday." This same technique can be used when cross-examining some witnesses. The witness can be directed to an either/or answer by including only two possibilities in the question.

With Mr. Spotte, however, I had no choice. I did not have any idea what the other dams at the Pittston operation were like. I assumed they must have been as carelessly built as the dam at Buffalo Creek, since Pittson invariably said the dam at Buffalo Creek was constructed in accordance with the customs of the industry. But I had been unable to obtain any information from other coal companies about their dams which might have enabled me at least to argue that Pittston's customary practices were reckless compared to all the other coal companies' practices.

So I asked Mr. Spotte about the other dams. As he testified I began to believe that the dam at Buffalo Creek was not built even in the customary Pittston way. First, Pittston's other dams had emergency spillways. Second, Pittston's other dams did not unnecessarily block a stream as well as impound black water from the cleaning plants. Third, Pittston's other dams had not been built as haphazardly as this one. The others had clay cores, or had been engineered in some way or another. The one at Buffalo Creek was merely refuse dumped out of the back of trucks into Middle Fork at a spot picked by the truckdrivers for their own convenience. Fourth, the dam at the Buffalo Creek operation was bigger, held more water, and was nearer to inhabited communities than any of the other Pittston dams.

After hearing this testimony, I decided to prove that Pittston's other dams had not been built as recklessly as this one. This would destroy Pittston's custom-and-usage argument, although it would give some support to Pittston's argument that it was a safe and careful company and did not know that the recently acquired Buffalo Mining Company was not in compliance with its normal standards. I attempted to deal with this secondary defense, lack of knowledge, in another way.

Mr. Camicia had already testified before the Ad Hoc Commission, as an aside, that Pittston had been able to buy the Buffalo Mining Company's stock because the then management of the Buffalo Mining Company felt it did not have the financial wherewithal to bring that company into compliance with the new Federal Coal Mine Health and Safety Act. I got Mr. Camicia to repeat this testimony during our depositions:

> [O]ne of the principal reasons that we were able to acquire this property was that the health and safety laws, the

> Federal Coal Mine Health and Safety Laws, had just been passed, as you know, effective April 1st of 1970, and they were so burdensome, so costly, that a company such as the then existing Buffalo Mining Company felt that they could not pay for these additional costs, felt that they did not have the staff or the know-how to even understand how to administer these things, and it gave us an opportunity to come in.

So, Pittston knew the company it was buying was not complying with federal safety standards and did not know how to comply. That's one of the reasons Pittston was able to buy it out. But after Pittston took over, it did not use its "staff" and "know-how" to bring the Buffalo Mining Company into compliance with the safety standards of the federal act. Indeed, despite Pittston's knowledge that the Buffalo Mining Company itself could not afford to bring its operations into compliance, Mr. Spotte admitted that he knew of not one single penny Pittston had made available to the Buffalo Mining Company in the two years it had owned it to enable it to make its operations comply with federal safety standards.

XV

"Nowhere Else to Put It"

It was time to question Steve Dasovich, the man directly in charge of the Buffalo Mining Company operation. I had heard quite a lot about him before this. He had a Spotte-type reputation. He was coming up the coal company ladder the same way that Mr. Spotte and many others before him had. He was tough, a tough superintendent and now a tough vice-president. He was known as a very hot-tempered fellow, one used to having his way around the mines. Nobody told Steve Dasovich what to do if he was working under him.

But the disaster apparently changed him. I was present when he testified at the Senate hearings three months after the disaster. By then he was very subdued, showing no anger and none of the tough-guy hostility I had heard he was famous for. He was a believable witness, not trying to excuse his conduct, but also not trying to explain it away.

> Senator Harold Hughes: In spite of being dumbfounded, and I am sure in a state of shock by it, how do you explain your failure to know what was taking place?

> Mr. Dasovich: Senator, I have pondered that question for 3 months now: What could have happened? I just don't know. I just don't know.

Although he appeared not to have any understanding as to how this disaster could have occurred, he didn't blame nature, as Mr. Spotte did. Mr. Dasovich accepted his part of the blame.

Steve Dasovich gave every evidence of having gone through the most horrifying mental suffering, thinking about the damage the dam had caused, the damage he had caused to his friends and neighbors. When he told me, during his deposition, what he saw from the top of the hill after the dam broke, he choked up and could not finish—

> I heard all those noises and my view was blocked from where I had turned in. I turned across the tracks. My vehicle drowned out. I got out and looked and went up on the tracks to see where all the noise was coming from and I looked up toward Lorado and it was throwing those big houses around like tenpins. Then I turned and looked towards Lundale and the town had already disappeared. The water struck that bridge just about where I lived and veered to the left and it took all the houses out. I didn't see anything down through there. . . .

Ironically, Mr. Dasovich might be a good witness for Pittston. How could a jury believe that this suffering human being could have acted with such reckless disregard as to have caused the death of over 125 of his friends and neighbors? If he was adamant and self-righteous, blaming nature or God, the jury might get angry with him. But they could hardly feel antagonistic to him when even I felt sympathy for him. At the trial I would have to keep Mr. Dasovich on the events that preceded the disaster, to show him as he was before the disaster, the Boss at the Buffalo Creek operation whom no one could question. I could not afford to let the jury see his other, human, side.

During his deposition, I took Mr. Dasovich back over the history of the Buffalo Mining Company's operation. Dam 1 had washed out during the rainy season in March 1967, before he started working for Buffalo Mining Company. But he learned about this failure after he joined the company. Maggie Rhodes could testify to that.

Maggie Rhodes had known Steve Dasovich all her life. She called him "Prejudice." He called her "Pride." In 1967 Maggie Rhodes—then Maggie Daniels—lived with her mother, father, sister, and four children in Saunders in the next to the last house on the left going toward the dam. She had lost her husband, Mr. Daniels, in a mine accident and had not yet married Mr. Rhodes.

"It was about ten on a cold night in March 1967. I was watching television with my sister. The rest of the family was asleep. The front door was shut. I heard someone scream. I ran to the porch. The creek was to its banks and was roaring. It was making so much noise I could not understand what the person was hollering, so I hollered back. The only word I heard the person hollering was the word 'dam.'

"I immediately got my family and the rest of the people and went toward the mountain. The water was in the road in front of our house. It was up to my waist as I ran, leading my children. They got soaked as we ran across the road in front of our house, through the yards of the houses across from us, then across the railroad track, and across the main road up onto the mountain."

"People were screaming, but there was no crying. We could see that the water looked fierce, because the church had on its dusk-to-dawn light. We finally got up the mountain a little ways and sat there for a while. McKinley Peters, who lived in the last house in Saunders, came by in a car. He could not get his car across the bridge to go downstream from Saunders, because the bridge was covered with water. So he took us in his car to the Saunders church. We stayed there awhile, until someone said the church would be the first thing to go. So we left there and went through the church to Ozzie Adkins's house where we spent the night.

"I did not realize until we got to Ozzie Adkins's house that I did not have my shoes on. I stayed up until around five A.M. From there I saw explosions, saw boulders flying, saw fire and steam. There was fear. It was just one of those nights.

"After that, I could not rest whenever it would cloud up to rain. When that would happen, I would take food and bed clothing and pack up the car and go and stay at Jean Cook's house in Lorado. I did this on several occasions, whenever there would be a rain or storm. We had three bridges to cross to get out of Saunders, and if there ever was another incident we would never be able to get out of there across that many bridges.

"Steve Dasovich knew I was afraid of the dams on Middle Fork. I had a conversation with him about that after he joined the Buffalo Mining Company. Steve just told me there was no danger. He just laughed at us, made fun of us. They were educated; they knew how wide it was supposed to be and what it would hold. We were just laymen, and we knew nothing about it. One of these times he even laughed at me about losing my shoes during the 1967 flood. He was laughing about me running barefoot through Ozzie Adkins's corn field. I did not think it was very funny."

During Mr. Dasovich's deposition, I asked him about this conversation with Maggie Rhodes. He denied it. I don't think he was lying. I believe he was too crushed by the disaster to lie. He may have completely forgotten about this conversation. I know that my sisters and I remember incidents from our childhood in different ways. Something that is important to me may have been completely unimportant to them and vice versa.

So too with this conversation. It may have been the most important conversation in Mrs. Rhodes's life because she had such a strong fear of the failure of the dams on Middle Fork. She was suffering as a survivor of the failure of Dam 1 and thus was acutely anxious and aware of rain and any other problems with respect to dams. But to Mr. Dasovich, who didn't live through the failure of Dam 1, and who did live with the fear of death in the mines all the time, this conversation probably was of little significance. For him, some lady asking about the safety of the dams on Middle Fork would be an unimportant event. So Mr. Dasovich may have been telling the truth, so far as he was capable of remembering it, when he denied this conversation with Maggie Rhodes had occurred.

But Mr. Dasovich did not deny that he had learned that Joe Holly, the inspector from the West Virginia Department of Natural Resources, a man concerned primarily with prohibiting black water pollution of streams such as Buffalo Creek, had written, after the failure of Dam 1 in 1967, that the then Buffalo Mining Company's management had been "warned numerous times of the danger" of a washout due to the lack of an emergency spillway for Dam 1. Joe Holly had written that strong action should be taken against the Buffalo Mining Company because of these prior warnings. Mr. Dasovich also knew that the state did not take any strong action. And he knew that the Buffalo Mining Company then put in a makeshift spillway for Dam 1 and that ended the matter.

By the time of the next rainy season—February 1968—Buffalo Mining Company had completed another dam behind Dam 1. This dam, Dam 2, was built because Dam 1 had begun to fill up with solid coal waste filtered out of the black water. But Dam 2 did not have an emergency spillway either, so it began to impound a large amount of water, as had Dam 1 the year before. This time the people in Saunders sought help. They had given up hope that the

Buffalo Mining Company would do anything to protect them, so Mrs. Pearl Woodrum, a Saunders resident who later became one of our plaintiffs as a result of the 1972 failure of Dam 3, wrote a letter to West Virginia's governor in February 1968. It was most prophetic:

> Dear Sir,
>
> I live 3 miles above Lorado. I'm writing you about a big dam of water above us. The coal co. has dumped a big pile of slate about 4 or 5 hundred feet high. The water behind it is about 400 feet deep and it is like a river. It is endangering our homes & lives. There are over 20 families here & they own their homes. Please send some one here to see the water & see how dangerous it is. Every time it rains it scares every one to death. We are all afraid we will be washed away & drowned. They just keep dumping slate and slush in the water and making it more dangerous every day.
>
> Please let me hear from you at once and please for God's sake have the dump and water destroyed. Our lives are in danger.

Mrs. Woodrum's letter brought state representatives to Middle Fork where they met with Mr. Dasovich on February 26, 1968—exactly four years to the day of the Buffalo Creek disaster. As a result of this meeting, Mr. Dasovich drew a one-page sketch for what he called a "new dam" to be built on Middle Fork behind Dam 2. This was Dam 3.

His sketch was not an engineering diagram. Just a little rectangle marked "new dam" drawn between some rough lines depicting Middle Fork.

> Q: Did you ever consult any engineers, civil engineers, or any books at all before you wrote those words "new dam" in connection with that design?
>
> A: No. No books.

Q: And no engineers?

A: No engineers.

* * *

Q: When you designed Dam 3 did you estimate the amount of acre feet [of water] that would flow into the pond behind Dam 3?

A: I did not.

I kept pressing Mr. Dasovich to admit he had not built Dam 3 in accordance with proper engineering practices. He finally told me what I wanted to hear.

Q: Now, in this book [the *Drainage Handbook* published by Mr. Holly's Department of Natural Resources] there is a discussion of how you calculate the minimum required principal spillway size. Do you recall reading that?

A: Yes.

Q: Had you made any calculations such as that to determine the size of the spillway that would be required for Dam 3 on Middle Fork?

A: There were no engineering calculations whatsoever on Middle Fork.

In May 1970 Pittston acquired all of Buffalo Mining Company's stock. Steve Dasovich's superiors, the two men who were the principal owners and managers of the Buffalo Mining Company, left the operation upon the sale. Pittston offered Dasovich the job of running the entire operation. He accepted.

Mr. Dasovich continued the work on Dam 3 after Pittston took over. Soon after Pittston bought Buffalo Mining Company's stock, Dam 3 reached the other side of Middle Fork Hollow and began to impound water. Then, during the rainy season in February 1971, almost the entire front, downstream side of Dam 3 slumped into the pool behind Dam 2. This was a major slide—150 to 200 feet wide, about one-third to one-half across the front of the dam, and 20 to 30 feet back into the dam. Despite this failure of Dam 3, exactly one year before Dam 3 was to give way completely, Mr. Dasovich and Pittston did not build an emergency spillway for Dam 3. I felt they had been warned to do so, by the failure of Dam 1 without such a spillway, by the damming up of water behind Dam 2 without such a spillway (which brought in the state inspectors), and by numerous inspection reports from Joe Holly, the West Virginia inspector who checked Dam 3 and wrote, on a number of occasions, that it lacked a "needed emergency spillway." Mr. Dasovich even admitted that Mr. Holly orally told him and the others at the operation that Dam 3 needed an emergency spillway.

How could Mr. Dasovich have ignored Mr. Holly's written and oral warnings? That is the coal industry in Logan County. Mr. Holly rarely was able to force Mr. Dasovich or any coal-company official to do anything. Why? He explained this during his testimony before the governor's Ad Hoc Commission.

> Q: In the matter of enforcement, do you ever run into any political pressure in Logan County, that you take it easy on the coal companies?
> A: Yes, sir.
> Q: You say you have?
> A: Yes, sir.
> Q: Have politicians come to you and said, "Joe . . ."—
> A: Not me.
> Q: Or go to the Division—
> A: They go further than me, I understand.
> Q: They go around you, then?
> A: You have to ask that question of somebody else.
> Q: But you think that does happen?
> A: There is no question about that. As I said, I think somebody is trying to run me off now.

Q: Do you catch this political pressure from both sides or just one side?

A: Both sides. Yes, sir. It is not confined to any one individual party.

There was additional evidence that Mr. Dasovich knew that dams need emergency spillways. In February 1972, the month which ended with the Buffalo Creek disaster, he had such a spillway built for a dam in the Elk Lick Hollow, the next hollow over from Middle Fork. The dam in Elk Lick Hollow was built with carefully calculated emergency spillways, in strict accordance with state regulations. Indeed, Mr. Dasovich even made sure that this dam was kept below fifteen feet in height to correspond with state law. Nevertheless, he allowed Dam 3 in Middle Fork Hollow to continue to increase in height beyond its already enormous sixty feet and built no emergency spillway for it. Dam 3 was just considered outside the law.

I asked Mr. Dasovich why he hadn't at least taken the trouble and time to have an emergency spillway built along the No. 5 mine road which ran up Middle Fork Hollow beside Dam 3. He had built such a spillway along the side of the dam in Elk Lick Hollow. He said there wasn't enough room in Middle Fork Hollow.

Q: If you blasted into the mountain you could make enough room.

A: And shut No. 5 mine down for a limited period of time.

Q: And built the spillway and opened the mine up.

A: Yes.

Q: What was so horrible about shutting down No. 5 mine?

A: It was a production unit. It affected about 90 men.

Q: You didn't want to stop production, is that right?

A: It's 90 men's livelihood, and getting them in and out of there.

That was the trade-off, jobs for lives.

I had also asked Mr. Dasovich how they could continue to dump refuse on Dam 3 after certain Federal Coal Mine Health and Safety Act standards prohibited the use of refuse piles to impound water. He said, "We had no other place to dump it."

I was struck by this response. During the depositions one of our lawyers went to the library in Charleston and checked out Richard Llewellyn's *How Green Was My Valley*. I did not remember much about this book, or the movie made from it, but I did recall that it depicted the devastation from coal mining on a valley in Wales, a valley no longer green after the coal company finished with it. In the book a young boy asks an old man why the coal company dumps all its coal-waste refuse in their valley:

> "Is the pit allowed to do this to us, Mr. Gruyffydd?" I asked him.
>
> "Do what, my son?" Mr. Gruyffydd asked.
>
> "Put slag by here," I said.
>
> "Nowhere else to put it, my son," he said.

XVI

"What About Existing Dams?"

Robert Reineke, the Pittston lawyer quoted in *The New Times* as having said that Pittston was responsible, was next. I hoped to get Mr. Reineke, under oath, to admit he made that statement. But he wouldn't admit it. Instead, he produced the letter he had sent to *The New York Times* denying he had said that Pittston was responsible. I did not tell him I had the reporter's notes of their conversation. If necessary, I planned to have the reporter testify at the trial, and Mose McGee-like, pull those notes from his pocket. But Mr. Reineke still was helpful.

Since he was Pittston's most senior corporate lawyer, I decided to ask him about two particular safety standards issued under the Federal Coal Mine Health and Safety Act. From the outset of this case, I had been trying to prove that, prior to the disaster, Pittston was violating, and knew it was violating, these specific safety standards. One standard provided that "refuse piles shall not be used to impede drainage or impound water." I felt Dam 3 violated this standard, since it was a refuse pile which impeded the drainage of Middle Fork, and it certainly impounded water. The other safety standard provided that "if a water or silt retaining dam will create a hazard, it shall be of substantial construction and shall be inspected at least once a week." I felt Dam 3 also violated this standard, since it retained silt, retained black water from the cleaning plant, and retained stream water. The disaster caused by its failure proved it created a hazard. It was not of substantial construction. And Pittston admitted it had not been properly inspected once a week.

On the other hand, Pittston argued that the safety standards were not violated, because Dam 3 was not a dam. They called it an "embankment," an "impoundment," or a "porous impoundment used as a water filtration system." As Mr. Camicia carefully explained to me, "We have never in the language of the coal field referred to that type of structure as a dam." And, they said, this embankment also was not a refuse pile. They said the only refuse pile was the dump along the side of the hollow, not the dump across the hollow.

I asked Mr. Reineke if anyone ever asked him for advice in connection with these particular safety standards. He looked pleadingly to Mr. Staker, apparently hoping Mr. Staker would protect him from the question. But the question was proper, so Mr. Staker told him to answer. Mr. Reineke remembered that one person, not connected with the Buffalo Mining Company operation, had asked his advice, "but I don't know whether it was before or after the disaster."

I pressed on. "Anyone else you can think of?"

He turned and talked with Mr. Staker, off the record and out of my hearing. He was still trying to avoid answering my questions, apparently on the ground that I might be invading a lawyer-client privilege. Mr. Staker told him he could "identify the person." Mr. Reineke then answered that "there was one other," but he gave me no name. No volunteering by Mr. Reineke.

"Who was that, sir?"

"D. S. Dasovich."

"When was that?"

"It was before" [the disaster].

This was an incredible answer, implicating Pittston directly in the disaster. This was testimony from Pittston's chief legal counsel that Mr. Dasovich had asked him for advice, before the disaster, about the specific federal safety standards which prohibited the use of refuse piles to impound water and prohibited the construction of retaining dams not of substantial construction. Although I had earlier asked for copies of any documents on this subject, Pittston had never produced any document to indicate that Mr. Dasovich had asked Pittston's lawyers about these standards. If there was such a document, it could prove conclusively that Pittston, rather than just Mr. Dasovich and Buffalo Mining Company, knew that Dam 3 was violating specific federal safety standards.

My job was twofold. I needed to get on the record as much as I could about Mr. Dasovich's request for advice. If possible, I wanted to prove there were some documents Pittston had not produced. Second, and more important, I needed to lead Mr. Reineke into testifying that Mr. Dasovich had not been asking for any legal advice from him, to ensure that Pittston could not claim any lawyer-client privilege with respect to any such documents. It would not be difficult to lead Mr. Reineke in this direction, since I assumed he would want to explain away any failure on his part to respond to whatever Mr. Dasovich had asked of him.

After much prodding and questioning, Mr. Reineke testified that Mr. Dasovich's request had come in a written memorandum in June of 1971. He refused to indicate exactly which safety standards Mr. Dasovich had asked him about, and then added, "I'd say these were questions, but this was really more of a memorandum with his comments on the standards as published May 22, 1971. And while there are question marks there, I did not take it really as an inquiry, but really as his comments."

The memorandum contained question marks, but Mr. Reineke was not about to admit that Mr. Dasovich had asked him a question. This gave me the opening I needed to establish that this was not a request for legal advice.

> Q: He wasn't really asking for your legal advice?
>
> A: He never received a reply if that's—that I recall.
>
> Q: Did you understand that he was asking for legal advice or not or are you telling me by not replying it was fairly clear he was not asking for legal advice?
>
> A: I did not feel he was.

I asked Mr. Staker for a copy of that document. Mr. Staker responded, "The fact of the matter is, unknown to Mr. Reineke, you have that document." Mr. Staker thought Mr. Reineke was talking about a document I already had. That may explain why he had not stepped in to protect his witness and Pittston from my discovering any more about this document. I quickly proved to Mr. Staker that I did not have this new document, so he said, "We'll take this matter under advisement." But it was too late. Mr. Reineke had already undercut any attempt Pittston might make to claim a lawyer-client privilege as to this memorandum. If the memorandum still existed, they would now have to produce it.

Pittston finally gave me a copy of the document, a June 2, 1971, memorandum from Mr. Dasovich to Mr. Reineke. It dealt with Section 77.215(e), the federal safety standard which prohibits the use of refuse piles to impede drainage or impound water. Nine months before the Buffalo Creek disaster Mr. Dasovich had written to Mr. Reineke, as Pittston's legal counsel, the following:

> Lorado plant pumps plant bleed water in Middle Fork Hollow. This hollow is blocked by a refuse constructed dam to allow the solids to settle out. The clean water percolates or is decanted. What about existing dams?

Mr. Reineke never responded.

Now we had written proof that Pittston's counsel had been notified, long before the disaster, that there was an "existing refuse constructed dam" on Middle Fork which was violating a specific federal safety standard. This memorandum was evidence of Pittston's knowing, continuing violation of the law.

Mr. Reineke also testified that he had been on a committee of the Bituminous Coal Operators Association (BCOA), placed there by Mr. Spotte as Pittston's representative, to provide coal-industry comments on these safety standards. I didn't know this before my questioning began. Indeed, I had not received any documents from Pittston showing that Mr. Reineke was involved with these safety standards, despite my request for any such documents back when the case first began. Mr. Reineke's testimony, though, caused Pittston to make a more careful search.

Within a few weeks after Mr. Reineke's deposition ended, Pittston sheepishly admitted that it had quite a large number of documents in its files relating to these safety standards. These new documents provided written proof that Pittston was even involved in the drafting of the very safety standards which it later knowingly and willfully violated.

The documents Pittston gave me still contained a number of gaps. I assumed there were additional documents somewhere showing more about Pittston's, and Mr. Reineke's, direct involvement in the drafting of these standards. So I called the BCOA in Washington and asked to see their files on these standards. They were in no mood to cooperate with my efforts in this lawsuit against Pittston. However, they knew I could get a subpoena for their documents, so they let me come over. I Xeroxed many of their documents and deposed Mr. Reineke, Mr. Spotte, and Mr. Dasovich again.

Mr. Reineke, on this second go-round, surprised me again. He identified some handwritten notes I had obtained from the BCOA files as his own notes. It turned out that Mr. Reineke personally made the oral argument to the Bureau of Mines when the coal operators made a last-ditch effort to strike the safety standard which prohibited the use of refuse piles to impound water. But Mr. Reineke's argument for Pittston and the BCOA failed. The Bureau of Mines made this a mandatory safety standard on May 22, 1971. And two weeks later, almost nine months before the Buffalo Creek disaster, Mr. Reineke learned from Mr. Dasovich that Pittston's "refuse constructed dam" on Middle Fork was violating this specific safety standard.

It would be many months before these revelations, and other such facts uncovered in our depositions of other Pittston people, became publicly known. Although Judge Hall had denied Pittston's attempt to seal the transcripts from public view, Pittston still had a right to delay the public filing of the transcripts while each Pittston witness reviewed his typed transcript and requested corrections in alleged mistranscriptions. The court reporter had to wait for Pittston to run out its time on these formalities before filing the transcripts.

I was happy to let Pittston take its time delaying the public filing of these transcripts. I hoped it would feel pressured by the forthcoming publication of these facts of recklessness we had uncovered. Maybe Pittston would try to settle with us now to forestall the public filing of these transcripts.

XVII

"Psychic Impairment"

We had finished the first round of depositions. We had established a pretty good case for reckless disregard out of the mouths of the Pittston people. By the time of the trial, if we had to go to trial, we hoped to buttress this case further with other Pittston admissions and documents, and with testimony from people in the Valley who could further implicate Pittston.

Meanwhile, we were on the defensive on the other front, responding to all Pittston's requests—interrogatories, depositions, medical examinations, even document requests. This took a massive army, since we could not afford to ask for extensions of time. Brad Butler, an Arnold & Porter lawyer who missed his natural calling as a general, led this army, which included, full-time, five other lawyers, four legal assistants, three law clerks, a document file clerk, and a document-room receptionist. Plus, from time to time, twenty more legal assistants and at least ten secretaries. Brad never had to ask for one day's extension of time despite Pittston's efforts to force us to seek delays.

While I spent my days and nights trying to find every possible fact to prove Pittston was reckless, Brad and our other lawyers dealt with the people. I slowly lost touch with them, seldom seeing them. I was involved with the procedural victories of the lawsuit, sometimes scoring in the depositions, forcing Pittston's witnesses to admit this fact or that fact, proving that Pittston was still holding back damaging documents. But whenever I felt any joy with these small victories, I became depressed. I felt better operating from a sense of rage and anger. Happiness seemed inappropriate. This should not be a fun lawsuit. Too many people had suffered, and were still suffering. Was I really advancing the real interests of the plaintiffs? Was our discovery helping to force a fair settlement, and fast?

After Brad and his team filed five books of interrogatory answers, Pittston began its depositions of the plaintiffs, four a day, every day of the week. This was the first taste of real litigation for many of our young lawyers. They would be battling the best litigators in West Virginia's largest law firm, trying to protect our clients from harassment and improper questions. This was also very harrowing work because they had to talk with each plaintiff at some length before each deposition. They had to sit in on the depositions of four plaintiffs each day, and then at night prepare the next four plaintiffs for the following day. That meant listening to eight very sad stories of emotional distress and death and destruction each day. One of our lawyers, Phil Nowak, moved to Charleston with his wife and child and went through this schedule day in and day out for five straight weeks. Eventually, though, we had to rotate the lawyers every week or two. The experience was too much to take for longer than that.

These depositions were even more difficult for the plaintiffs. Many of them felt uneasy going to Charleston to spend the night at a hotel. Some didn't feel that they dressed right or looked right. Often they would not leave their rooms to go out to a restaurant for dinner. They'd order up hamburgers or something else from room service. We paid for these room-service bills, and for their rooms, and let them keep the $20 plus travel expenses Judge Hall had ordered that Pittston pay each of them for the trip to Charleston. This way the people at least got the satisfaction of a check from Pittston for their Charleston ordeal.

During the depositions, Pittston's lawyers tried to undermine the plaintiffs' mental-suffering claims, if that was at all possible. They tried to show, through very carefully drafted questions, that the plaintiffs had not suffered any physical injuries and therefore could not have suffered any mental injuries. This is a legal question which later was to become a central issue in the case. But we never argued that the plaintiffs' mental injuries were caused by any physical injuries. Indeed, we admitted that most of them had not been physically touched by the water. Nevertheless, Pittston insisted on demonstrating over and over again that the plaintiffs had suffered no physical injuries.

> Q: Now, Mrs. Wilburn, did any of the water from the flood waters or the trash and debris on the flood waters ever come against your body, hit against your body or cause an impact with your body?
> A: No, sir.
> Q: Did you receive any physical injury to your person or to your body from the flood waters or the trash and debris in the flood waters?
> A: No, sir.
> Q: Did the waters, that is, the flood waters of Saturday, February 26, 1972, ever come against, hit against, strike against, or cause an impact on the bodies of any of your children?
> A: No, sir.
> Q: Did any of your children receive any physical injury from the flood waters or the trash and debris in the flood waters, that is, an injury to their body or person?

> A: Not as I know of, no.
> Q: Was there any evidence upon your body of an injury to your body or person?
> A: No, sir.
> Q: Did you receive any lacerations, cuts, scratches, bruises, sprain or strains of muscles, fracture of bones or swelling or anything of this nature?
> A: No, sir.
> Q: Did any of your children receive any lacerations, cuts, scratches, bruises, swelling, sprain or strains of muscles or fracture of bones?
> A: No, sir.

But even Pittston's lawyers eventually stumbled on the fact that you didn't have to have lacerations to suffer mental injury.

> Q: Mrs. Wilburn, did you on the railroad track get to a place of safety away from the flood waters?
> A: Yes, sir, but we didn't know if the water was going to raise up higher and come to the railroad tracks or not.
> Q: Now did you stay there at that place?
> A: Well I stayed there until I got sick and nauseated and started vomiting and I couldn't stand to watch, I watched all of it, but I got deathly sick at my stomach, and it seemed like I just wanted to pass out.

We also asked the plaintiffs some questions during their depositions, after Pittston's lawyers had finished. Ordinarily a lawyer doesn't ask questions of his own clients at a deposition. But I wanted the depositions to show more than the answers to Pittston's carefully worded questions. I thought Judge Hall or his law clerk, Stanley Dadisman, might be reading these depositions, and I wanted to be sure they knew the whole story. Pittston, of course, objected to our fouling up their carefully crafted examinations. For example, we asked Wanda Wilburn a number of questions after Pittston's counsel had finished. Not surprisingly, Pittston's counsel objected.

> Q [by Ken Letzler, plaintiff's counsel]: Have you had trouble sleeping since the flood?
> Pittston's counsel: Show our objection.
> Plaintiff's counsel: Can you answer the question?
> A: Yes, I have.
> Pittston's counsel: Move to strike the question.
> Plaintiff's counsel: Are you bothered by rain?
> Pittston's counsel: Show our objection.
> Witness: Yes, when it rains, I can hardly go to bed at night.
> Pittston's counsel: Move to strike the question and answer. . . . I am saying that this is a discovery deposition, and that the questions which you are asking are not proper on a discovery deposition. Your client may talk to you and tell you whatever she wants to at any time. [But not here.]

Our questions usually dealt with sleep problems, nightmares, fears, and other manifestations of anxiety. We began to ask these questions after two little children told us of their nightmares. One little boy dreamed he was caught under the water. He woke up in a cold sweat, holding his breath and gasping for air. On the day of the disaster he had not been in the water at all. He had seen everything from the hill. In his dream, though, he vividly relived the ordeal of those who were in the water. He was so frightened by this dream that he never told his brothers or sisters or parents about it.

There were numerous such dreams. I was still a little doubtful whether any of our psychiatric experts would be able to persuade a West Virginia jury about their survival-syndrome theories. But I thought a jury might understand the horror of the disaster through the retelling of the nightmares the people were still living with. These nightmares seemed to be the most concrete type of evidence available at the time, at least until we could get our own psychiatrists to see each one of the plaintiffs.

Eventually Pittston's counsel began asking the questions they knew we would ask, about the troubles the plaintiffs might have had since the disaster, about their dreams or nightmares, and about any fears whenever it rains. These questions were upsetting to the plaintiffs. Often they would break into tears, and become hysterical, as they retold their stories of the disaster. Pittston's counsel soon became embarrassed by this and asked us, off the record, to indicate to them when they were getting into a touchy area so they could desist from asking any upsetting questions. We refused. We felt Pittston's counsel should sit through this emotional trauma. Let them cringe and feel how difficult it would be for them if they took this case to trial. If Pittston's lawyers were having trouble listening to these stories, they might begin to imagine how a jury would react.

Pittston's counsel tried in a number of ways to get the plaintiffs to testify that they were not suffering from any mental injury. We had chosen the words "psychic impairment" to describe their mental suffering. We thought that people in West Virginia would not willingly state on the public record that they were suffering any mental illness, and we hoped that "psychic impairment" would be a more neutral-sounding term. But the plaintiffs would not even admit they had a thing called "psychic impairment."

So whenever Pittston's counsel asked a plaintiff if he or she suffered any "psychic impairment," we objected. We argued this was a legal term. We said Pittston's counsel could ask all the factual questions they wanted about the plaintiffs' fears of rain, nightmares, nervousness, and the like, but they could not ask a legal question. Judge Hall agreed with us that Pittston should not trick the plaintiffs into giving up their claims by asking them if they had "psychic impairment." Nevertheless, some of Pittston's lawyers continued to do this. So we continued to direct the plaintiffs not to answer, without running back to complain to Judge Hall.

Then in August, Pittston filed a lengthy document request asking the plaintiffs to produce any and all documents showing how and why they had hired Arnold & Porter. Coincidentally, within two weeks after this Pittston document request, the West Virginia State Bar decided to rekindle its long-dormant investigation of Arnold & Porter. The *Gazette* headlined the news—"Bar to Examine Soliciting Client Cases." Private investigators hired by the state bar went to the Valley and began questioning over a hundred plaintiffs as to how and why they retained Arnold & Porter. When Charlie Cowan called and told me the investigators were coming to question him, I advised him to tell them he would only talk to them if his counsel was present. The bar's investigators skipped him.

Later, though, when Pittston served notice of its intention to take Charlie Cowan's deposition in our lawsuit, I decided I'd better show up personally. I guessed Mr. Staker might try to use this deposition to ask the leader of the Citizens Committee how Arnold & Porter was selected.

I guessed right. Even though Mr. Staker had not questioned any of the other plaintiffs (that was done by the Charleston law firm specially hired by Pittston for that task), he was there personally to question Charlie Cowan. Not surprisingly, he soon asked Mr. Cowan how and why the people had retained Arnold & Porter. I objected and instructed him not to answer the questions.

He also asked Mr. Cowan if he was suffering any "psychic impairment." I objected, and Mr. Staker offered one of his most delightful arguments: "I have not asked if he has a claim for psychic impairment. I am asking him if he claims to have suffered an impairment of the psyche. . . ."

Soon thereafter we got a chance to raise a number of these issues before Judge Hall. I waited until Pittston set a date for a hearing on another matter, and then filed a document with Judge Hall setting forth a series of items which I thought should be discussed at Pittston's hearing. High on our list of significant items were Pittston's questions of the plaintiffs about their retention of Arnold & Porter and about psychic impairment.

Judge Hall again said that Pittston's questions on psychic impairment would do Pittston no good. "Harassing these witnesses by making them admit over and over that they don't have some disease or that they don't know what it means . . . is not really going to help the defendant in the defense of this case. I'll just tell you . . . you are going to be wasting your time because when we get to the trial of the case, the Court is not going to let you ask that kind of a question when you are cross-examining some witnesses, so you are engaging in futility. . . . I think . . . when you get over into these questions on top of questions and make the person admit that they don't have something that they don't know the definition for is unfair to the witness and shouldn't be done. I think the only way I can say it is to reiterate the admonitions earlier given that you shouldn't do that."

Judge Hall proved once again that he understood what was going on in this case. He felt for the people. He did not want to let them unknowingly give up claims which their lawyers felt they legitimately had, and he would help us protect them.

As for the bar's investigation, I told Judge Hall that I thought there was a concerted effort to separate the plaintiffs from their lawyers. I said I had no objection to the investigation by the West Virginia State Bar, but I did object to some of Pittston's lawyers using the depositions in our case to harass the plaintiffs about the hiring of their counsel.

I added that the plaintiffs "have made a major effort in this case to hold out against great odds. They are people without money, having faced a great disaster, having decided they don't want to settle with Pittston for the kind of settlements that Pittston has made. They want to do this thing legitimately, above board and with strength, and then to come over here and have themselves be questioned about how did you hire your lawyer, what right did you have to go and hire those lawyers, that's just not proper and I do hope there isn't any need for a ruling on that, your Honor, but it has come up during these questions, and I did put it in the memorandum."

Judge Hall told Pittston not to ask any further questions of the plaintiffs about how they hired their lawyers. He said that was a matter for the bar association.

We raised one other important matter with Judge Hall. By the time of this hearing, Pittston's medical experts had examined approximately 300 plaintiffs at Williamson, Kentucky—men, women and even young children. Unlike the depositions at Charleston, at which the children under six years of age were not deposed, clients of all ages were medically examined by Pittston's doctors. But we had not been provided with a copy of any of the medical reports on any of these 300 or so plaintiffs.

I wanted to know what was happening at Williamson, especially since we were not allowed to sit with them during their examinations. We knew that Pittston's chief doctor, Dr. Russell Meyers, was sixty-nine years old and a board-certified neurologist. There was also a psychologist, Dale Stanton. We assumed he too was a doctor. Dr. Meyers's wife, a registered nurse, and various other medical attendants also were involved in the examinations. To have as complete a record as possible, assuming as we did that we were going to have great trouble from the outset in getting Pittston to turn over to us all the materials produced in these examinations, we decided to debrief each of the plaintiffs as they left Williamson.

We hired a history teacher from the Man Senior High School to talk to each of the plaintiffs after the tests were completed. Her very complete reports provided us with detailed information about what Dr. Meyers and his people were doing. This would help us later in cross-examining Dr. Meyers, especially if he left out some of the information the people remembered telling him.

Each plaintiff had to undergo a physical examination, stripping naked to do so. They also had to provide a complete medical history and take a series of psychological tests administered by Dale Stanton.

There were five major psychological tests. First, there was a standard sentence-completion test. The plaintiff was asked to complete sentences beginning with words such as "I hate . . ." and "I suffer. . . ." Sentence-completion tests such as these sometimes provide significant insights about an individual. However, at the same time the sudden confrontation of the patient with his own attitudes may be extremely threatening and therefore may lead to his use of trite or facetious completions, often of little diagnostic value. Nevertheless, this test and the rest were legitimate tests. A question would arise only if we disagreed with Pittston's interpretation of the test data.

The second test was a figure-drawing test. The plaintiffs were asked to draw various pictures—man, woman, house, and tree. Often a person reveals a great deal about himself and his attitude toward others in the way he draws and elaborates these pictures. For example, if the human figure is drawn small and over to the side of the paper, all alone, this may indicate withdrawal. If the human figure is in the middle of the paper, but lopsided and with no stable ground drawn underneath, this may connote feelings of insecurity.

The third test was the Bender-Gestalt Test. Each plaintiff had to reproduce a set of nine designs (including a row of dots and a row of adjacent circles) on a sheet of blank paper. Certain diagnostic groups perform this test in a rather consistent way. For example, mental defectives, when producing the line of dots, often will continue the row until it goes off the page. Similarly, many schizophrenics tend to squeeze their reproductions into the smallest possible available space and arbitrarily rotate the designs. This test is useful for some of the symptoms common to the plaintiffs, such as depression (small designs) and anxiety (much needless erasing and retracing).

The fourth test was the Thematic Apperception Test. It consists of a series of cards with pictures of men, women, and children in various situations; for instance, a child staring sensitively at a violin, a man with an angry expression pulling away from a woman who is holding his shoulders and attempting to restrain him. Each plaintiff had to interpret the picture, to tell the psychologist something about what was going on in the situation, what led up to it, how the characters were feeling, and how it might turn out. These stories might reveal the emotional needs of the plaintiff and his attitudes toward significant figures and situations in his environment. There is a corresponding Children's Apperception Test which uses ten cards with animals instead of people, although the animals are in humanlike situations. It did not appear from our debriefings that Pittston's psychologist knew about the Children's Apperception Test, since he was using the adult cards for the children.

Finally, there was the psychiatric interview. Dr. Meyers questioned the plaintiffs about the disaster and their concerns since the disaster. He also administered a galvanic skin-response test to the plaintiffs by connecting a series of wires from the patient's fingers to a meter-type instrument, often identified by the plaintiffs as "looking like a lie detector." Indeed, this test is often administered as part of a thorough lie-detecting-test procedure. Dr. Meyers was using it to register the patient's level of anxiety to specific stimuli after he first determined an initial level of response to "neutral" stimuli. Various words, phrases, and familiar names, as well as pictures of the Buffalo Creek disaster and of other floods, were used by Dr. Meyers as stimuli.

Although the plaintiffs were intimidated by the galvanic skin test and by other aspects of these day-long medical examinations, many still described Dr. Meyers as a kindly, gray-haired doctor. I assumed he was a West Virginian who had spent many years at the Appalachian Regional Hospital. I was greatly concerned that a West Virginia jury would find this Appalachian doctor more credible than our Ivy League doctors. But then I learned that Dr. Meyers, who was born in Brooklyn, had come to the West Virginia area only in 1963. He had been chairman of the Division of Neurosurgery at the University of Iowa Hospitals for almost twenty years before he unexpectedly and precipitously left Iowa for the hospital in Williamson, Kentucky. His expertise was in surgery, primarily neurosurgery, rather than psychiatry. Since we were not complaining of physical or organic diseases, but of mental diseases, it appeared that Dr. Meyers's qualifications were not very useful for the task which Pittston had given to him. Later, when Mr. Staker described Dr. Meyers to me as a "hard, dictatorial, taskmaster," I felt more relieved. If the jury saw him as hard, and not old and kindly, we'd be better off.

We could not find any information on Pittston's psychologist, Dale Stanton. He was not listed in any of the American Psychological Association directories. Eventually we learned that he was only a psychologist-in-training, not qualified yet to refer to himself as a psychologist. Mr. Staker had not known that Dale Stanton was not a doctor, and not even a psychologist, since Dr. Meyers had hired him. When Mr. Staker did learn this, he told the court he was not sure Pittston would have Dale Stanton testify at the trial.

Even though we'd gone to a lot of trouble to find out what was happening at Williamson, we still wanted copies of Pittston's actual medical reports. If Pittston's doctors were finding widespread evidence of severe mental suffering, then we might be able to force a settlement which would include payment for such suffering without spending the money to hire our own psychiatrists to individually examine each of the 625 plaintiffs.

So we asked Judge Hall to order Pittston to produce the medical reports which Dr. Meyers and Dale Stanton had prepared for Pittston. Pittston said they would produce Dr. Meyers's reports only if we would produce copies of our medical reports. Judge Hall agreed with Pittston and ordered an exchange of medical reports. He also ordered Pittston to include Dale Stanton's reports in the exchange, despite Pittson's protest that his reports were not medical reports required to be produced, now that they had learned he wasn't a doctor.

We had obtained individual psychiatric reports, from Dr. Lifton and some psychiatrists from the University of Cincinnati Medical School, on a sampled group of fifty plaintiffs. However, if I wanted all Pittston's medical reports, I now would have to get our own psychiatric reports on all the rest of the plaintiffs. This would be very expensive and would force the plaintiffs to undergo another set of medical examinations. We decided not to go ahead with this yet. Maybe we could settle the entire lawsuit before the deadline for exchanging medical reports.

Meanwhile, we learned again how lucky we were to be in federal court before Judge Hall, instead of in state court. During this summer, while the depositions and medical examinations were proceeding, the papers in Logan County, and in Charleston, were filled with stories of a mine-machinery theft ring and massive arrests of Logan County people by the state police. Judge Oakley, the Logan County judge who had presided over the grand jury's determination that no one should be charged with the murder of 125 people in the Buffalo Creek disaster, also presided over the criminal proceedings involving these people who had stolen from coal companies.

The first story on this ring reported that Judge Oakley

had sentenced a forty-year-old man who had pleaded guilty to receiving $250 worth of stolen auger bits to one year in the county jail. Judge Oakley indicated he had been lenient with this man because he had cooperated with law enforcement officers in the investigation.

XVIII

"Danger Note"

Pittston had delayed the filing of the transcripts of our first depositions of their people as long as possible, making suggestions for word changes here and there. Finally, however, there were no more excuses for holding up the signing of the transcripts. After Pittston's people signed them, the court reporter filed them with the court clerk's office, where they were available to the public.

The *Charleston Gazette*'s Mike White wasted no time. In his first story, based on Mr. Camicia's deposition, he wrote that Pittston no longer termed this disaster an act of God. Mike quoted Mr. Camicia as saying, "The impoundment was man-made and machine-made, and, of course, it was the cause of the disaster." The *Charleston Daily Mail* headlined its story on this admission, "Pittston Head Acknowledges Buffalo Dam Caused Flood."

The headline on Mike's next story was "Danger Note to Buffalo Delayed by Pittston Head." This article, which appeared on the first page of the *Charleston Gazette*, was based on our earlier discovery depositions of Mr. Spotte and Don Jones.

Don Jones was in charge of all of Pittston's preparation plants. He was one of the twenty-one people we deposed in the first go-round. He had testified that he prepared a memorandum, over a year before the Buffalo Creek disaster, to "notify all of our people concerned with the preparation plants of the proposed [federal safety] regulations which would affect them." In this memorandum he specifically warned that these regulations would "forbid the closing off of any stream or the impoundment of water" by refuse-pile dams.

After preparing this memorandum, Mr. Jones went in to clear it with Mr. Spotte. This was normal procedure. When Mr. Spotte saw what Mr. Jones had written, he ordered him to collect all the copies of the memorandum and not to send them out. Mr. Jones did as he was ordered. He went "back across the street after the conversation," took the letters "out of the envelopes and put them back in the files," and threw away the envelopes—in December 1970, over one year before the Buffalo Creek disaster. This was the "danger note delayed by Pittston head."

These memoranda lay in Pittston's files, undelivered, until two days *after* the Buffalo Creek disaster. Then, on February 28, 1972, when Mr. Jones learned about the disaster, he went to his files, pulled out the letters of warning, and sent them to all Pittston's preparation-plant managers. This time he didn't wait for Mr. Spotte's approval. Mr. Jones would at least forestall any further disaster.

This was not the only time important materials which should have been sent to the people at the preparation plants were held up at Pittston's headquarters. We also learned during the first round of depositions that in July 1971, seven months before the Buffalo Creek disaster, Clark Todd, the Pittston Coal Group's vice-president of industrial engineering and training, received a booklet prepared in England by the National Coal Board to ensure against another Aberfan disaster. The introductory paragraph in this booklet, *Spoil Heaps and Lagoons*, gave the simple warning, "All spoil heaps [refuse piles] and lagoons [black water settling ponds] should be so constructed and drained as to prevent the undue accumulation of water within them."

Mr. Todd gave this copy of *Spoil Heaps and Lagoons* to Mr. Spotte. Mr. Spotte read it and passed it on to Mr. Jones. Mr. Jones then asked Mr. Todd if he could make copies of it for distribution to Pittston's preparation-plant managers. Mr. Todd "questioned the propriety of doing so without permission and the matter was dropped."

Why did he question the propriety?

Mr. Todd said he was concerned about some kind of copyright problem, and "It was a National Coal Board study which had really been a gift to me from an acquaintance who held a responsible position on the coal board and I didn't know really, as I said, the propriety of a distribution on our part."

Did Pittston write to the National Coal Board and ask for permission to circulate it? No. It just put it in its files. That is, until the Buffalo Creek disaster. Within five days after this disaster, Mr. Jones had numerous copies of *Spoil Heaps and Lagoons* printed by the Pittston Coal Group printing shop. They were distributed to all the Pittston preparation-plant managers and engineers. This time no one bothered about any copyright problem.

The stories in the *Wall Street Journal* also helped educate the public about this disaster. The *Journal* had covered the disaster when it occurred. At about the time Mike White's stories on the depositions were appearing in the *Gazette*, the *Journal* reporter went back to the Valley for an update and wrote a front-page story on "Survival Syndrome." This story emphasized that mental anguish and guilt generally afflict victims of disasters, and it quoted from doctors who'd been seeing this phenomenon at the Logan-Mingo County Mental Health Clinic, set up after the Buffalo Creek disaster. I hoped Pittston's shareholders, and its management, soon would begin to understand the significance of our mental-suffering claims, now that the *Wall Street Journal* had put its imprimatur on "survival syndrome."

The national television news also added to the pressure on Pittston. ABC-TV's Jim Kincaid had arrived from Chicago soon after the disaster. He tried to go up the Valley to photograph the remains of the dam, but his film crew was turned back by the state police. They finally found a back way by an old log-hauling road and worked their way up to the dam site. At the last minute, a national guardsman stopped them. He asked if they were the ABC film crew. Kincaid, resignedly, told the truth. The guardsman said, "Well, I was sent to stop you." He then looked down the hill to the Valley, lying in ruins, and said, "Go ahead. My family lived down there."

Kincaid continued to report on the disaster and its aftermath for ABC. His reports had helped to force Governor Moore to create the Ad Hoc Commission and kept the story of Pittston's poor safety record before the public. Later, ABC began a series of special investigative reports, with Kincaid's report, "ABC News Close-Up on West Virginia—Life, Liberty and the Pursuit of Coal," as the first. This report aired in prime time on ABC national news in October 1973. It told of the fact that Pittston had been assessed over $2 million in fines since the federal government passed the 1969 Coal Mine Health and Safety law, but the Bureau of Mines records showed Pittston had paid none of these fines. ABC even found that, over a year and a half after the Buffalo Creek disaster, "in hollow after hollow, small towns and villages sit directly in the path of waters dammed up by huge piles of coal slag dumped by neighboring coal operations." Some coal companies still gambled with the people's lives.

This was an important message for the people in West Virginia, some of whom might one day sit on the jury in our case. One day when I was flying back from Charleston to Washington, a young man struck up a conversation with me. He was with the West Virginia Chamber of Commerce. I told him I was working on the Buffalo Creek case. He said, "I feel real sorry for the people in the Valley. They have really suffered, and I hope you have a good outcome in your case." But speaking as an employee of the West Virginia Chamber of Commerce, he was afraid we might get too large a verdict against the Pittston Company. "This would take money out of the pockets of West Virginia and its coal companies."

I had to explain to him that a large verdict against Pittston in favor of the survivors of Buffalo Creek would be putting money into the pockets of West Virginians. I also had to remind him that the money for the verdict would be coming from the Pittston profits which now flow outside the state to its shareholders. These profits go to people like Mr. J. P. Routh, the chairman of the board of the Pittston Company and its largest stockholder. Mr. Routh probably hasn't been in West Virginia in a generation. He maintains one residence in New Jersey but spends most of his time at his home in Florida. Mr. Camicia, Pittston's president, lives at a fine home on Long Island, in Locust Valley, New York. Mr. Kebblish lives in Pittsburgh, Pennsylvania. Mr. Spotte lives in Dante, Virginia. Taking money out of their pockets would not be taking money out of the pockets of West Virginians.

The fact is that West Virginia is not owned by West Virginians. It is a colony, owned and controlled by absentee landlords. A West Virginia newspaper study in 1974 demonstrated this.

> [M]ore than two-thirds of the non-public land in the state is controlled by outside interests—giant fuel, transportation

> and lumber companies. Often paying low property taxes, they extract the state's rich deposits of coal, timber, oil and gas. And their activities inevitably help sustain the striking paradox of a state with abundant mineral wealth and much abject poverty.

Logan County's statistics tell the story. Ten companies owned 220,494 acres of Logan County's 291,725 privately owned acres.

XIX

"Enough"

On a number of occasions I had drafted a motion to ask Judge Hall to speed up the setting of a trial date. But each time I decided not to file it and instead to wait for a more opportune moment. Pittston had finally finished all its medical examinations and was about to finish its depositions when Mr. Staker asked me if he could have some additional time beyond the court's present schedule to review all these materials. I refused and assumed he would ask Judge Hall to order this extra time. We had opposed these kinds of delays all along. But in the past, each time that Pittston had asked for more time, Judge Hall had agreed.

We were reaching the second anniversary of the Buffalo Creek disaster. I couldn't take another delay, and I doubted the people could either. So I prepared a short memorandum to Judge Hall opposing "any further delays which may be requested by the defendant." I wrote in anger, and it read that way. I circulated the memorandum to some of my partners at Arnold & Porter. They objected to its tone and told me it should be more lawyerlike. But I'd already mailed it to Judge Hall by the time I got their comments. I hoped it would not antagonize him. This is what I wrote, in part:

> The defendant has been given more than ample time to complete its discovery. And the defendant has made full use of its discovery rights. Indeed, there is hardly a rule unused. Almost every plaintiff has been required to travel to Charleston for depositions. Almost every plaintiff has been required to travel to South Williamson, Kentucky, for medical examinations and a battery of psychological tests. Every plaintiff has had to search for and produce a mass of documents, and answer a host of interrogatories as well as sign numerous authorization requests. And some of the plaintiffs have been required to permit inspections of their homes or trailers.
>
> Enough. Already two years have passed since the Buffalo Creek disaster. The second anniversary will occur on February 26. Let us end this interminable discovery and get on to the trial of this case.

Lawyers object to one-word sentences, especially a one-word exhortation like "enough." That's not a lawyer's sentence. It doesn't have a verb, it doesn't have some of the other things that sentences are supposed to have. And what's worse, it shows anger, emotion. And the words "hardly a rule unused." That's much too flip.

We appeared before Judge Hall on February 15, 1974, almost two years after the Buffalo Creek disaster, one and a half years after we had filed our complaint. A lot of time had gone by, over 600 plaintiffs had been deposed and medically examined. After disposing of a number of preliminary matters, Judge Hall, apparently having read our memorandum, asked Mr. Staker, "What do you have to say about the extra time on discovery?" Mr. Staker began a response which the judge cut off with "Do you want some more time or not?" Mr. Staker finally asked for at least four months more time to complete discovery. Judge Hall said, "All right. I am going to rule on this right now."

I jumped up, afraid he was going to grant another delay. I asked him to hear our argument. I said I did not understand why Pittston needed four more months for discovery and made an impassioned, anxious speech trying to stop this further delay. Judge Hall let me finish, but from his ruling I realized he had already made up his mind. "All right. This case is set for trial on July 15."

I was stunned. That was only five months off, an early trial date for a case this size. I hadn't had the nerve to ask for a trial date in our memorandum. All I was asking for was an end to any further delays to complete discovery, knowing there were a number of steps still to come even after discovery ended before Judge Hall would be ready to set a trial date. He had indicated earlier that he would not decide on a trial date until all these steps were completed, so I never expected him to react to our memorandum by short-circuiting everything to set a trial date.

I was often astonished by our good fortune during this case. Bud Shay once said, "I see the hand of the Lord operating here. Things have been happening in this case which are just too unusual. We aren't this good." I, too, think the hand of the Lord was present. Our lawyering didn't bring Judge Hall into this case. In fact, had we been able to keep Judge Christie, the original judge, our problems on February 15, 1974, would have been overwhelming. We learned that morning that Judge Christie had just died. Had we still been before him, our case would have been delayed until a new judge could familiarize himself with the cumbersome record now on file. Instead, we were before Judge Hall, and he had just set our case for trial.

And when Judge Hall said July 15, he meant it, too. He looked down at us from the bench and said, slowly and deliberately, "I am setting this down for a trial date, and we are going to trial then. That's what we are going to do."

Judge Hall filled in the rest of the schedule for the remaining months between this February 15 hearing and the July 15 trial date. He extended the deadline for discovery from April 1 to May 1 and told each side to get all their medical reports completed by then if they intended to use any medical testimony. This extension to May 1 actually benefited us more than Pittston. Pittston had already completed all its medical reports, but we had not really begun. This extension also gave us more time to depose some of Pittston's people. There were still some loose ends to tie up, and we intended to do that.

Judge Hall told us he planned to have a trial on four to six representative cases, and asked us to see if we could agree which cases should be tried first. He also gave each side until May 1 to work out any fact stipulations, to list all their exhibits, and to file all their pretrial motions. He set June 7 for a pretrial hearing and June 15 for the filing of a trial brief by each side. Everything was set.

Pittston, looking for some way to delay matters, asked us to indicate to Judge Hall how long we thought it would take to try the corporate-veil issue. Its lawyers said they thought maybe a month or two. I said that issue could be dispensed with quickly based on the admissions we already had from the depositions. Of course, Pittston disagreed. Judge Hall had the last word, "We will take however long is necessary."

PART THREE

XX

"Ambulance Chasing"

I was ecstatic with the progress of our lawsuit. But before we could fully celebrate our good fortune, an old problem cropped up again. When we got back to Washington, I found a letter on my desk from James W. St. Clair, the new chairman of the West Virginia Bar Committee on Unlawful Practices, inviting me to meet with them on March 9 "to discuss with the members of your firm the matters surrounding your representation of a great number of people in the Buffalo Creek, West Virginia, area."

We knew about Mr. St. Clair, a lawyer in Huntington, West Virginia. A year and a half before this, when we filed our lawsuit in his home town of Huntington, he wrote, angrily, to his fellow members of the Committee on Unlawful Practices that Arnold & Porter should be investigated for ambulance chasing.

Ambulance chasing describes a lawyer who rushes to the scene of an accident, following the ambulance siren, to solicit the dazed victim to hire him as the lawyer. This is unethical. Actually, though, lawyers are not so crass in our big cities. Instead, a lawyer may pay off policemen, ambulance drivers, tow truck operators, hospital attendants, or others to distribute the lawyer's card to the injured accident victim. It is also unethical for a lawyer to solicit business this way. Arnold & Porter had not engaged in ambulance chasing. We'd been called to the Valley by the survivors' committee, and the Supreme Court had held, almost ten years before this, that accepting such a committee's call for help was not "ambulance chasing."

Maybe Mr. St. Clair didn't know all the facts surrounding our agreement to represent the survivors. Or maybe there was some other reason for his concern. It is not unheard of, in American history, for lawyers to use their state bar associations to try to keep out competition, or even to prohibit needy persons from obtaining necessary legal advice. For example, since 1883 the Brotherhood of Railroad Trainmen has assisted in the prosecution of claims by injured railroad workers or by the families of workers killed on the job. The Brotherhood has done this by recommending to its members and their families the names of lawyers whom it believes to be honest and competent. In the early 1960s, the Virginia Bar Association and the Virginia state court determined that this arrangement constituted an improper solicitation of legal business and the unauthorized practice of law by the Brotherhood of Railroad Trainmen. In 1964, the United States Supreme Court disagreed. The Court understood what was at issue:

> Injured workers or their families often fall prey on the one hand to persuasive claims adjusters eager to gain a quick and cheap settlement for their railroad employers, or on the other to lawyers either not competent to try these lawsuits against the able and experienced railroad counsel or too willing to settle a case for a quick dollar.

The Supreme Court then held:

> It is not "ambulance chasing." The railroad workers, by recommending competent lawyers to each other, obviously are not themselves engaging in the practice of law, nor are they or the lawyers whom they select parties to any soliciting of business.

Ten years later, in 1974, it appeared that Mr. St. Clair, and his committee, might not have gotten the message which the Supreme Court had given loud and clear. Our representation of the survivors of the Buffalo Creek disaster fit the situation set forth by the Supreme Court in the Virginia Railroad Trainmen case. Very soon after the Buffalo Creek disaster, many survivors did "fall prey" to the claims adjusters who went around to their homes and made repeated efforts to get them to settle their claims against the Pittston Company.

Further, there were lawyers who were not competent to try their cases against the coal company's able counsel, and there were lawyers too willing to settle their cases for a quick dollar. There were some lawsuits filed for some of the other survivors, but as far as I know, most of these cases were quickly settled without any attempt by those survivors' lawyers to put Pittston's feet to the fire through extensive discovery, one of the best weapons available to plaintiffs to force a defendant to make a fair and equitable settlement.

So the Buffalo Creek Citizens Committee acted properly in recommending counsel to the survivors, and Arnold & Porter acted properly in responding to the committee's request for help. When we met with Mr. St. Clair and his Unlawful Practices Committee we explained that to them. I also said that I understood they had been in contact with Pittston's counsel in conducting their investigation. Some members of the committee quickly denied this, so I informed them that Mr. Staker told Judge Hall, in open court, that he had given the bar committee a list of plaintiffs to be questioned by the bar's investigators. One member of the committee had to agree that had happened. Apparently, the other members did not know about this. This was certainly some indication that the bar committee might, unknowingly perhaps, have been helping to further Pittston's efforts to separate these survivors from their lawfully retained counsel.

That ended the session. They said they would meet and make a report to their bar association. Already, though, they had taken up too much of our time, diverting us from our major job of representing our clients.

XXI

The Statute of Limitations

We were nearing February 26, 1974, the second anniversary of the Buffalo Creek disaster, and the last day of West Virginia's two-year statute of limitations. Generally speaking, West Virginia law required that any lawsuit against the Pittston Company for damages caused by the Buffalo Creek disaster be filed within two years.

We faced a difficult moral dilemma. When we filed our complaint in September 1972, we represented only about 400 of the 4,000 or more survivors. Because of the publicity of our lawsuit, about 200 additional survivors later asked to join in. Many in this second group had already settled with Pittston for their property losses or for the wrongful deaths of members of their families, but we felt they still could sue for their own psychic impairment. We had amended the complaint to add these 200 to the lawsuit, and Pittston had immediately used that amendment as an excuse for extending the time for completion of its depositions of the plaintiffs. This had delayed the case for the original 400 plaintiffs, many of whom had not yet recovered anything for their own property losses because they had sued Pittston rather than settle. It was not right to ask the original 400 to sacrifice any further, so we decided not to add any more plaintiffs after that.

Just before the running of the two-year statute of limitations, some of our original plaintiffs asked us to make an exception to our policy of turning down any more cases, even if this might delay their own cases. They asked me to add Sylvia Jean Davis, a young girl only sixteen at the time of the disaster. They said, "She's really had it bad, and she hasn't gotten anything yet." I agreed to talk to her.

She told her story as though it had occurred only two weeks before instead of two years.

"The sky was dark and there was some lightning at that time. All at once water started coming into the house through the back kitchen door. I grabbed my eight-year-old brother and my two-year-old sister in my arms and tried to go out the back door. But the water was too deep, and the house started to float. I then tried to get out another door of the house and climbed onto the roof of one of our cars that was parked there. But the house started to move again so me and my brother climbed off the car and jumped back into the house with my mother and sister. I was holding my sister on my back. My mother was holding my brother. The house began to float again. Then a burst of water came into the house and swept us out the door into the rushing water. I was caught in the middle of logs and a great deal of other debris. This was the last I saw of my mother and my eight-year-old brother. I was trying to keep me and my sister afloat. I made it for a few minutes. But then I was banged into a tree and my sister fell off my back. I can still recall her scream as I saw her go floating away. I never saw her again."

She remembered her tortuous path literally from log to log, from debris to debris, in just the same way that Dennis

Prince could not forget his struggle, frame by frame, second by second. She was thrown around and mauled by the raging torrent filled with debris. Eventually she was rescued, although she suffered numerous bruises and cuts and has a scar two inches by one-half inch on her leg.

Her father had settled with Pittston for the wrongful deaths of her mother, her eight-year-old brother and her two-year-old sister, and for the family's property loss. As an heir, Sylvia would share in these payments, but she recovered nothing for her own damages, her own mental and physical injuries. Pittston rarely made any payments to the children for their own losses. We added Sylvia Jean Davis as a plaintiff on the very last day before the statute of limitations ran.

The state of West Virginia also filed a lawsuit against Pittston, just before the running of the statute of limitations, for the state's damaged or destroyed bridges, roads, and schools. The state asked for $100 million in damages—$50 million for compensatory damages and $50 million for punitive damages. But the state did not try our gamble. It brought two lawsuits, one in federal court and an identical backup suit in state court.

The federal government also had a right to sue Pittston for its damages, which were considerable. The United States spent at least $2 million immediately after the disaster to clean up the Valley and rechannel Buffalo Creek. Its total expenditures, including the moneys paid for bringing in the HUD trailers, and for providing disaster loans, educational facilities, free food stamps, and emergency unemployment compensation to those left jobless by the disaster, totaled another $7 million. Much of these funds came from the President's Disaster Relief Fund, after President Nixon, from China, declared the Buffalo Creek disaster a natural disaster.

On the eve of the running of the two-year statute of limitations, a number of newspaper reporters called the Justice Department in Washington to see whether the United States would be filing a lawsuit against Pittston. They were told that no such suit was planned. When Congressman Ken Hechler of West Virginia learned this, he sent telegrams of protest to the Attorney General and the Federal Disaster Assistance Administration expressing his "shock at the report that the Federal Government through the Justice Department was not going to seek reimbursement for federal expenditures resulting from the disaster." Congressman Hechler summed it up: "It is outrageous that taxpayers should have to foot the bill for this disaster."

But that is exactly what happened. The federal government brought no lawsuit against Pittston. All of us as taxpayers helped bail out the Pittston Company by paying millions of dollars to mitigate the damages caused by their disaster.

This second anniversary of the disaster also triggered nightmares and fears as the disaster victims remembered and relived again their experiences from the day of the disaster. To help alleviate some of these fears, I wanted to be able to reassure the people that Pittston was not building another dam on Middle Fork. I obtained permission from Pittston's lawyers to take a trip up to the old dam site.

I arrived and began the long trek up Middle Fork Hollow on a Saturday morning. The mines weren't working. No one was around and it was quiet. Pittston knew I was there, but they made certain that no one from Pittston joined me. I guessed they wanted to be sure I couldn't question Pittston people without their counsel being available. As I climbed up the hollow, past the still-smoldering remains of the refuse pile, along the bubbling waters of Middle Fork's stream, I kept thinking about *Deliverance*, fearing a sniper's shot would ring out and I would be no more. I was paranoid, maybe because this hollow had killed so many people already, maybe because it was nearing the second anniversary and I was experiencing the same kind of increased fear and heightened awareness as the disaster victims. Whatever the reason, I was terrified as I climbed alone, thousands of feet up this long hollow.

I had been at work on this case for over a year and a half and still had no real feeling for what Middle Fork Hollow looked like before the dams gave way. By helicopter I think I might have begun to understand the magnitude of the dams. From ground level, it was beyond my comprehension. After over an hour's steady climb up the hollow, I decided that Pittston was not building another dam on Middle Fork. But there were so many more such hollows in the area. I was in no position to assure anyone that Pittston wasn't building another dam in the next hollow.

I did confirm the fact that there was an old dam on Middle Fork further up the hollow from where Dams 1, 2, and 3 had been. This dam did not give way at the time of the Buffalo Creek disaster and was still intact when I saw it some two years later. A small spillway had been cut in the rock around this dam when it was built. That is why it did not fail when the other dams on Middle Fork, without spillways, did. This was proof, if any further proof was needed, that the failure of Dam 3 was not an act of God. The rainfall that preceded the disaster was not so heavy or so excessive as to cause this dam on Middle Fork to fail, so the rain would not have been too much for Dam 3 if it had been properly built, with an emergency spillway.

That evening I went down into the Valley and met with the people at the Buffalo Creek Grade School. I wanted to discuss settlement with them, especially since Judge Hall told us, after he had set the trial date, that now might be a good time for the parties to begin settlement discussions.

I told the plaintiffs we had Pittston on the run. There would be a few more months of discovery, further harassment by Pittston, probably more interrogatories. But we had our trial date, and the clock was ticking away for Pittston. Now we could begin serious settlement discussions from a position of strength. I thought Pittston would insist that I obtain full authority to negotiate a binding settlement for all the plaintiffs before we could begin discussing settlement numbers. So I asked the plaintiffs to give me the authority to enter into, and complete, settlement negotiations on behalf of all of them, with authority to settle each case with Pittston for whatever individual figure I decided was fair.

After I explained all this at the meeting, Claude Rogers, one of the plaintiffs, stood up and made a motion—"If I am in order, I'd like to make a motion that we authorize Mr. Stern to settle all our cases for what he thinks is fair." Someone quickly seconded the motion, and it passed unanimously. I was moved by this show of confidence in us. I had never met Mr. Rogers, but I made it a point to talk with him after the meeting. He had a little book in which he collected autographs from the lawyers, law clerks, legal assistants, secretaries, doctors, psychologists, nurses, what-have-you, from everyone working for him and his friends on their cases. We meant a lot to him, and he wanted to show it. So he thought up the idea of making a motion.

Brad Butler told me of another incident that occurred during this visit to the Valley that demonstrated the faith the plaintiffs had in us.

While Brad was visiting Mrs. Nora Kennedy, one of our clients, he overheard her call two of her pigs "Arnold and Porter." One of the other plaintiffs, Mrs. Doris Mullins, was taken aback by this and told her that wasn't a very nice thing to say. Mrs. Kennedy responded that she meant no offense. "I call the pigs Arnold and Porter because they're rooting for us."

When I got back to Washington to draft a letter to all the plaintiffs confirming the authority they had given to me, I learned that the canons of ethics prohibited me from going ahead with my plan. A lawyer representing multiple plaintiffs may not settle any of the cases of any of the plaintiffs unless he first informs all the plaintiffs of the total amount of the settlement, and of the amount that each of the plaintiffs will receive, and then obtains approval from each in writing. This meant I could not insist that the plaintiffs agree, beforehand, to be bound by my division of any total settlement I might agree on with Pittston.

So we modified our letter to the people to indicate that they had authorized me only (1) to *conduct* (not complete) negotiations for a settlement on behalf of all the plaintiffs on the basis of one dollar figure for the entire group and (2) to prepare a *recommendation* (not a binding commitment) to each individual plaintiff as to the amount that each should receive from the total figure.

I was ready to talk settlement.

XXII

A Written Settlement Proposal

With the authority I now had from the plaintiffs, I sat down to draft a written settlement proposal to send to Pittston. I wanted our proposal in writing so there would be a better chance that our position, in our own words, would go directly to Pittston's management. I assumed Mr. Staker would report to Pittston's management on our proposal, but if I made our proposal orally, he'd probably summarize it in his words. If I presented it in writing, he might have no choice but to forward a copy to Pittston.

Drafting the written proposal was difficult. I wanted to begin serious settlement discussions, so I knew our proposal had to be credible. I couldn't just ask for $64 million. Everyone knew that we didn't expect to recover that amount. But I also knew that any proposal we now made would become the highest amount we could ever hope for. So I didn't want this new maximum to be too low.

A more difficult question was what was the minimum figure we would accept as a fair settlement for the plaintiffs.

Brad calculated that our plaintiffs lost about 100 homes, and that a solid, provable damage figure for the real and personal property losses of all 600 or so plaintiffs would be about $2.5 million, not including any amount we might recover for psychic impairment. But that didn't tell me what to propose to Pittston.

Many of us felt that $8 million, or maybe even $6 million, would be a substantial victory. One guideline that lawyers sometimes use for pain and suffering in automobile accident cases is a multiple of provable losses. Eight million dollars would net $5.5 million for psychic impairment, over twice our provable property losses of $2.5 million. Six million dollars would provide $3.5 million for psychic impairment, almost one and a half times our provable property damages.

Bud Shay, an experienced negligence lawyer, pointed out for me how difficult it is to determine what would be a good settlement for us. For example, if you think your case is worth $2.5 million, but the defendant would have gone as high as $4 million, then $2.5 million isn't a good settlement. But you never know how far the defendant will go, so $2.5 million, from your plaintiff's point of view, may be a good settlement. However, you also have to consider the jury. If the jury would give you $8 million, then $2.5 million is not a good settlement, and neither is the $4 million the defendant might be willing to pay. On the other hand, if the jury would only give $1 million, $1 million is a good settlement, even though it is less than you think your case is worth. But at the time when you are talking settlement, you don't know the jury would give you $8 million, then $2.5 million is not a mum value on the cases and try and get as much above that as you can."

In this case, a good settlement would have to provide enough money to each plaintiff to build and furnish a new home and clothe a family, with enough left over to put into savings for future medical bills and education for the children. And the total settlement figure also would have to be large enough to make the people feel they had beaten Pittston.

While I was struggling with this settlement question, I got a call from the office of Maine's attorney general. They had decided to help Maine's environmentalists in their efforts to stop Pittston from obtaining a permit to build an oil refinery at Eastport, Maine. They wanted some estimate on the amount of damages Pittston might have to pay our plaintiffs, so they could determine whether Pittston had the financial ability also to pay for any damages which might occur if a tanker broke in two in the waters off Eastport.

Their call reminded me that a good settlement might also have to include some payment to the plaintiffs for removing the problem which their lawsuit posed in Pittston's bid for its Maine refinery. If we settled before trial, Pittston would avoid the adverse publicity of a public trial as well as any jury finding that Pittston was responsible for the Buffalo Creek disaster. So, a settlement before trial in West Virginia might help Pittston in Maine.

In fact, although I knew it wasn't right to think this way, I even began to wonder whether a good settlement also should include some payment to me for giving up the right to try the case. I knew there were some major disadvantages to the plaintiffs in having to go through a lengthy and harrowing trial. Still, I couldn't help dreaming about the personal publicity I'd get as the lead counsel in a trial which would be covered daily by the national news media. Maybe the TV news would even carry an artist's sketch of the "survivors' lawyer." I had just read an article in *The New York Times* on "traveling lawyers," such as Clarence Darrow, who, down through the years, had traveled to various parts of the country to try famous cases. Clarence Darrow had been my hero from youth. It was his life story which started me thinking about becoming a lawyer. Reading this article made me aware of how much I wanted to try this case just so I could play Clarence Darrow. I tried not to let these thoughts divert me from my responsibilities to the plaintiffs, but it wasn't always easy.

I also realized I was reluctant to end the case because I then would have to return to a more normal legal practice. I doubted I would ever find another case as meaningful to me as this one. It had become my life, filling all my waking hours with excitement and purpose. I suspected I would lose this exhilarating feeling about the practice of law once I had to give up Buffalo Creek.

With some hesitation, I finished our settlement proposal. I wrote that we would settle for a total of $32.5 million as compensatory damages for all real and personal property losses, wrongful deaths, and psychic impairment. I didn't tell Pittston how I calculated this figure. Nothing scientific. One-half of $64 million, with $500,000 added on to make it appear that it wasn't merely one-half of $64 million.

I added that we would be willing to forgo recovery of punitive damages if Pittston would concede that it was responsible for this disaster. I was not forgoing much. Since Pittston's insurance policy did not cover punitive damages, Pittston could not agree to a settlement that provided for any payment of punitive damages. Thus, if we wanted to settle with Pittston, we would have to settle only for compensatory damages. But our settlement proposal for compensatory damages, since it included almost $30 million for psychic impairment, had an element of punitive damages in it.

We also agreed to limit substantially our request for injunctive relief if the case could be resolved by settlement. In our complaint we had requested an injunction to prohibit Pittston from continuing its past practices with respect to refuse piles and dams at all Pittston's mining operations in West Virginia, Kentucky, and Virginia. In our settlement offer we proposed that the injunctive relief be limited only to Pittston's operations in Logan County, West Virginia.

Since our lawsuit was based on the failure of Pittston's Logan County dam, that was about as broad an injunction as we could really hope for.

I'd agonized over our settlement proposal for weeks. Once it was mailed, I was able to forget it. It was out of my hands, and I could get back to work on proving Pittston's recklessness.

XXIII

"Flood at Lick Fork"

Some of the people I'd listed for depositions no longer were working for Pittston. This meant I could talk with them without Pittston's counsel being there.

One such person was Jack Morris, Pittston's former chief engineer. He had retired before the Buffalo Creek disaster. Mr. Morris's name appeared on numerous documents, since he had built many of the refuse-pile dams at the other Pittston locations. His engineering plans and memoranda indicated that he had never built a dam as recklessly or haphazardly as the dam which Steve Dasovich built at the Buffalo Mining Company.

Jack Morris was loyal to the Pittston Company. He had worked for them for many years and remained a Pittston stockholder. He still lived in Dante, Virginia, where the Pittston Coal Group headquarters was then located. And many of his friends still worked for Pittston. But he was also a man of great conscience. He was troubled by the Buffalo Creek disaster.

We talked about each of the other Pittston dams. I showed him diagrams we had obtained from Pittston and asked him how each was constructed—trying to get him to confirm that he had carefully engineered this one, that one had a rock spillway, that another had a clay core, that it would be reckless not to have an emergency spillway, and so on. I flattered him on how careful he'd been when he was Pittston's chief engineer, that it was a shame he'd had to retire before seeing the Buffalo Creek dam. Surely if he had seen it he would have tried to correct its obvious mistakes.

After more than an hour of this, Mr. Morris suddenly blurted out, "There is something I could tell you." But then he said, "No, I shouldn't tell you about that." I went on as though he hadn't said anything and continued to talk about these other Pittston dams. Then I mentioned, in passing, that I thought I knew what he was thinking about telling me. I referred to something, I can't remember what. He said, "No, that wasn't it." I asked where this thing took place, hoping to keep him talking on the subject. Finally, he said, "Another Pittston dam failed a while back."

He said this dam failure had occurred many years earlier, and he could not recall any of the details. He did remember, though, that it was reported at the time in the local papers.

As soon as we got back to Washington we began the search for these papers. Eventually we located a copy of the *Dickensonian*, the local paper in Clintwood, Virginia. Its headline for March 17, 1955, was, "Dam Break Causes Flood at Lick Fork." The story read like a slow-motion preview of the Buffalo Creek disaster some twenty years later.

> The great wall of water, approximately forty feet high at the tipple, roared down the narrow creek bed, sweeping two houses away. One of these was the residence of Mr. and Mrs. Bertsie Stanley and their three children.

* * *

> The black waters tore open the door, and caught the tiny baby, a few weeks old, in a whirlpool from which the mother snatched it and placed it atop a sofa. By that time the house was floating down the creek with the swift current. It lodged against a tree some 200 yards down the stream from which point miners formed a human chain and rescued them. The house was partially collapsed.

After dozens of phone calls, we found Mrs. Stanley. She told us she had hired a lawyer and filed a lawsuit, some twenty years ago, against Pittston's Clinchfield Coal Company. But, she said, the lawsuit went badly, so she finally agreed to settle for $10,000, much less than their actual damages.

I immediately made a document request on Pittston for all documents relating to this dam failure. They found some, but not all, of the papers on the Stanleys' lawsuit. The documents showed that Pittston's lawyers had tried the same defenses, without success, that they later raised in our case: act of God and custom and usage. First, act of God—Pittston argued then that "the raining was of such prolonged length and intensity that defendant avers that it constituted an act of God as the precipitation was greater than had been previously known at this place." Second, custom and usage—Pittston argued then that the Clinchfield "dam was properly constructed and consistent with the general customs and practices of similar dams." Pittston's lawyers must have known that act of God and custom and usage would not shield Pittston from liability, even twenty years ago, because they did agree to settle with the Stanleys.

More important, I now had proof that Pittston had direct corporate knowledge, long before the Buffalo Creek disaster, of the dangers of its own black water dams.

XXIV

"The Potential for Similar Disasters"

I also persisted in my efforts to prove that Pittston should have changed its refuse-pile practices because of the Aberfan disaster. In addition to the newspapers, magazines, and television accounts that told of the Aberfan disaster, there was an English tribunal's report on Aberfan which should have alerted Pittston to the dangers of refuse piles. After seventy-six days of hearings in 1966 and 1967, this tribunal issued its report in July 1967. This Aberfan report was placed on sale to the public for fifteen shillings—then around three dollars. Pittston never bothered to buy a copy.

Mr. Camicia had agreed, during his deposition, that Pittston should have obtained a copy of this report. He could safely say that, since at the time the report came out, he was not working for Pittston. He was working for a coal company in Illinois, an area of the country he described as flat, "as contrasted to the problem in West Virginia of no land surface or flat surface." So Mr. Camicia, in flat Illinois, wasn't concerned about the Aberfan tribunal's report.

But he had admitted that "if I had been located in West Virginia, at that time, I would have to assume that I would be interested in having a copy of it."

But even after Mr. Camicia became Pittston's president, and the Buffalo Creek disaster occurred, Pittston showed no interest in the Aberfan report. Indeed, even after we alleged in our complaint that Pittston was reckless for ignoring the lessons which should have been learned from the Aberfan disaster, Pittston's lawyers continued to treat Aberfan as irrelevant. Mr. Staker consistently objected to my questions about Aberfan. He argued that "the Aberfan Report and conclusions to be drawn from it are extraneous to matters concerned with the Buffalo Creek disaster." He added, "I object to and reject out of hand counsel's self-serving statement that there is a lesson from Aberfan to be learned in connection with Buffalo."

However, by the time I'd finished questioning Mr. Camicia, I was sure he was beginning to see Aberfan's relevance. He asked us if we had an extra copy of the Aberfan report. They decided it was time to read it—six years after Aberfan, fifteen months after Buffalo Creek, and eight months after we filed our lawsuit.

I had also questioned Mr. Spotte about his knowledge of Aberfan. He never saw the Aberfan tribunal's report, but he did remember that "Mr. Round, who is my friend and an employee of the National Coal Board," had sent him a report on Aberfan "published by the National Coal Board." This was a different report from the *Spoil Heaps and Lagoons* booklet prepared by the National Coal Board after Aberfan.

This report had not been turned over to us in response to our document request. Mr. Spotte explained why. "It was a personal copy. I read it and I don't recall what I did with it afterwards." He had no idea where it was now.

We immediately wrote the National Coal Board in England to get a copy of their report on Aberfan. The National Coal Board, which runs the nationalized coal mines in England, wrote back that there was no National Coal Board report on Aberfan. They referred us to the Aberfan tribunal's extensive report. But Mr. Spotte had been very specific in his deposition testimony that his friend, Mr. Round, had sent him a National Coal Board report, not the tribunal's report.

We wrote back to the National Coal Board that the report we wanted was sent by a Mr. Round to Mr. Spotte. We asked if they'd contact Mr. Round, whose first name we still did not know, to see what that report was. They replied that they would if we would pay their expenses. We immediately agreed.

The next letter from the National Coal Board was a gem. It quoted at length from a letter which Mr. Round wrote to them in connection with our inquiry:

> Such information and detail as sent [by me to Mr. Spotte] was prompted by a very long and fruitful friendship, reciprocal exchanges, and more especially *by virtue of the fact Irvin Spotte was operating a number of situations which had the potential for similar disasters*, as he was disposing washery discard and effluent on a number of mountain sides. [Emphasis added.]

This was a bombshell, direct evidence that Mr. Spotte knew, or should have known, that he was "operating a number of situations which had the potential" for another Aberfan disaster.

We wrote back and asked the National Coal Board to send us a copy of Mr. Round's letter, but they refused. So all we had was their letter quoting Mr. Round's letter. This National Coal Board letter would not be admissible as evidence at the trial because it was hearsay. There was no proof that the National Coal Board accurately quoted Mr. Round. I needed Mr. Round to authenticate either his own letter or the quotes attributed to him in the National Coal Board's letter.

But Mr. Round was in England and our courts have no power to order him to come to the United States to testify. I assumed he would voluntarily come and testify only if he could find some way to explain away his letter so as not to damage his friend, Mr. Spotte. Mr. Round probably wrote his letter to his old employer, the National Coal Board, because he thought he was answering a Pittston inquiry. His letter, as quoted by the National Coal Board, began, "There appears to be some mis-understanding in connection with the Pittston's legal enquiry." I doubted he would have been so open if he had known the inquiry was from plaintiffs who were suing Pittston.

Maybe I could get Mr. Spotte to admit the facts in Mr. Round's letter. I deposed Mr. Spotte again. During this deposition, I asked him a number of questions using the facts stated by Mr. Round. Mr. Spotte denied them all. I now had no choice but to get Mr. Round's letter authenticated.

There is a little-used procedure in our Federal Rules of Civil Procedure, a Letter Rogatory, which permits a deposition of someone in a foreign country. A Letter Rogatory is a formal letter of questions forwarded by our State Department to a foreign government's state department, requesting that government to arrange to have a particular person in their country answer the questions. It is a wonderful-sounding document—"The President of the United States of America to the Appropriate Judicial Authority in the United Kingdom, Greetings."

I had to apply to Judge Hall to get a Letter Rogatory. I told him, and the opposing side, the questions I wanted asked of Mr. Round and attached a copy of the letter which the National Coal Board had sent us. Even if I couldn't get the Letter Rogatory, or if Mr. Round later explained away his letter, at least I had used the document to tell Judge Hall what Mr. Spotte's friend, Mr. Round, had said. Judge Hall quickly issued the Letter Rogatory.

This became public when Mike White found the Letter Rogatory in the court's formal papers and wrote an article for the *Gazette*—"Briton's Testimony Sought in Dam Suits." Now the people in Charleston also knew that Mr. Round thought Mr. Spotte "was operating a number of situations which had the potential for similar disasters" to Aberfan.

I sent Judge Hall's signed Letter Rogatory, with our attached questions, to the Office of Special Consular Services at our State Department. They sent it "to the appropriate Foreign Service post for presentation to the judicial authorities of the country through diplomatic channels." Fortunately it is easier to do this with the United Kingdom than with non-English-speaking foreign countries, since translations are not required. The American Embassy in London then forwarded our Letter Rogatory to the United Kingdom's Foreign Service, which forwarded it to the High Court of England. They appointed a legal officer to ask Mr. Round our few questions. I just wanted Mr. Round to authenticate the quotes attributed to him in the National Coal Board's letter, so I could show that letter to the jury, and so I could use that letter in cross-examining Mr. Spotte at the trial.

We hired an English solicitor to shepherd our Letter Rogatory through the English High Court. In England lawyers are divided into solicitors and barristers. Solicitors deal with clients but rarely appear in court. They are like our office lawyers, advising clients on legal problems—wills, divorce, business problems, taxes, corporate mergers, things like that. Barristers, on the other hand, usually don't talk to the clients, except through solicitors. They only get into the act once the case gets to court. They are like our trial lawyers and appellate advocates.

When it came time for the actual questioning of Mr.

Round, our solicitor suggested we also retain a barrister to protect our interests. We did so. Pittston did too. Eventually our questions and Pittston's cross-questions were asked, and Mr. Round's answers were taken down in writing. As I guessed, Mr. Round tried to explain away his letter. For example:

> I was under the impression that the Pittston lawyers . . . were asking these questions . . . and the result was I naturally thought they would have all the information at their disposal and I did not give the attention to the letter, because I did not really know what the letter was for. I did not give the letter I would normally have given to them had I been called upon—knew what the purpose of the letter was for.

But we had already scored with Mr. Round. Judge Hall knew what Mr. Round had said before the lawyers got to him. And maybe some of the jurors in Charleston now knew about Mr. Round's warnings to Mr. Spotte from the *Gazette* story.

XXV

"Catch-22"

Ten days after we mailed our settlement proposal, Zane called. By now we were on a first-name basis. "I'm calling about your letter. I would have gotten back to you sooner, but my law partner died. I want you to know we are willing to meet to talk settlement, as we have always been willing to meet. But I note the $32.5 million you mention in your letter. We don't have $32.5 million in mind. Still, we'd be glad to meet to discuss this matter with you, especially since the court told the parties to see if they could reach some kind of settlement."

"I'm ready to meet with you at your convenience."

"Well, I want to talk to Mr. O'Farrell [Pittston's Charleston law firm] and Mr. Murdock [Pittston's New York law firm] to arrange a time. The sooner the better. I'll get back to you."

I was happy. Zane always insisted that the plaintiffs had to make the first proposal. I had, and he hadn't completely laughed it off. Apparently they had some figure in mind and would give it to us once a meeting could be arranged.

But it was obvious that these settlement discussions were not moving rapidly. So we had to proceed with a psychiatric examination of each plaintiff to be in a position to exchange written medical reports on each plaintiff by Judge Hall's May 1 deadline.

This would cost over $100,000. My partners at Arnold & Porter looked long and hard at our request that they advance that much cash. This was different from the $1 million in lawyers' time we had already run up. That wasn't an out-of-pocket cash payment. They agreed, though.

We hired the University of Cincinnati Medical School's Department of Psychiatry. The university's medical school is the nearest major one to this part of Appalachia, and a few of its psychiatrists, including Dr. James Titchener, Dr. Frederick Kapp, and Dr. Janet Newman, had been helping us for some time. They sent over sixty psychiatrists, social workers, and psychologists on two trips of two days each to see over 600 people in the Valley. We were going to do in four days what Pittston's doctors at Williamson, Kentucky, had been doing for over six months.

Just scheduling the doctors and the plaintiffs for these two-day trips was a horrendous task for our legal assistants, but they accomplished it. The sixty professionals were divided into teams. Each team interviewed one family at a time, first as a family unit and then each individual family member alone.

The initial team interviews with the family began with a request that the family talk about their experiences during the disaster and thereafter. Dr. Titchener remembers that in nearly every home this technique was "akin to setting a psychological fire, as the family's recalling of their experiences reactivated vivid memories and intense feelings from the time of the disaster two years before."

Psychiatrists are used to dealing with troubled people. But the psychiatrists who went to the Valley told us they were more shaken by their experience with these people than they had been in any of their other psychiatric work. A few of these psychiatrists were survivors of Nazi concentration camps who later moved to this country to practice medicine. Some also had patients who were survivors of Nazi concentration camps. So the reaction of these doctors, who spent only four days in the Valley, was itself symptomatic of the emotional effects of this disaster. Afterward the psychiatrists were terribly depressed, and it took quite a while to get them to sit down and write a report for each individual they had seen. This was another example of psychic numbing. The psychiatrists were overwhelmed by these disturbing tales which they themselves witnessed only secondhand through the stories of the survivors two years after the disaster.

In those families with children, the psychiatrists also obtained from the mothers a brief outline of each child's early developmental history and the child's functioning in major areas before and after the disaster. This information was passed on to the child psychiatrist who would then see each of the children separately.

The children were seen in their own rooms or spaces in the trailers. They were given the opportunity to recall their own experiences of the disaster. In many cases they had never spoken of their feelings about the disaster before. The child psychiatrists found pronounced effects of the disaster on the children based on the child's own level of development at the time of the disaster, the child's perceptions of his or her own family's subsequent reactions to the disaster, and the child's direct exposure to incidents at the time of the disaster.

The fact that the Buffalo Creek community was so completely destroyed caused serious problems for the survivors for months and years thereafter. As a result, the parents' reactions and the parents' inabilities to reconstruct their own lives had long-range effects on the children, despite the fact that the children might not have even seen the disaster, or been much troubled by it at the time it occurred. This finding was consistent with that of Anna Freud, Sigmund Freud's daughter, when she studied children who survived the bombing of England during World War II. She learned that children were more upset by their own mother's fears than they were by the bombs exploding around them. Thus, typically, a mother's inability to reconstruct her life after the Buffalo Creek disaster may have been a major cause of the subsequent anxieties which her children began to develop after the disaster.

Had the Pittston Company made efforts to reconstruct the Valley soon after the disaster, admitted its own wrongdoing, and helped the parents—for example, by funding a mental-health effort—some children might have been saved from serious emotional difficulties. Of course, many of the children were disturbed by the disaster, not only because of the subsequent anxieties of their own parents, but also because the children themselves witnessed ghastly scenes during the disaster at a very important developmental point in their lives.

The individual psychiatric interviews confirmed what we had already learned from Dr. Lifton and our sample psychiatric interviews—no one who survived the Buffalo Creek disaster escaped emotional and mental distress as a result of it. Our psychiatrists individually interviewed 613 plaintiffs. There were only 9 plaintiffs who demonstrated no degree of psychic impairment; 115 plaintiffs demonstrated mild psychic impairment; 301 demonstrated moderate psychic impairment; 182 demonstrated severe psychic impairment; and 6 were psychologically incapacitated. These ratings of psychological impairment were based on (1) psychological symptoms (such as anxiety, depression, phobia, other mood disorders, sexual dysfunctions, and sleep disturbances); (2) somatized, or what laymen would call physical, disturbances (headaches, gastrointestinal complaints, backaches, fatigue, and other bodily complaints with a psychological basis); and (3) personality problems (alcoholic addiction, developmental deviation). The severity of the psychic impairment was determined, in part, by an estimate of the persistence and entrenchment of the symptoms or personality change.

Pittston's doctor also found widespread evidence of mental disorders. But he generally refused to attribute these disorders to the disaster. Instead, he developed a "Catch-22" explanation. He decided that almost everyone in the disaster suffered a "gross stress reaction" to the disaster resulting in a "transient situational disturbance." Transient situational disturbance is the psychiatric definition of "shell shock," often found in soldiers. By its own terms, it is "transient," it goes away quickly.

But Dr. Meyers was seeing the survivors some eighteen months after the disaster. So, since he generally found no "perceptible improvement" in their mental condition by this time, their mental disorders could not be classified as "transient situational disturbances." Dr. Meyers had an answer: "That such improvement has not occurred impels the conclusion that other psychodynamic factors anteceded. . . ."

In other words, you should be cured of your mental disorder by the time you see Dr. Meyers. If you are, the disaster caused you only a minor disturbance, a transient thing, for which Pittston would have to pay only minimal damages. But if you are not cured, if you continue to evidence severe mental suffering eighteen months later, you must be suffering from a "preexisting vulnerability" or mental condition. In that case, under the law, Pittston would be responsible only for the aggravation of that condition.

Dr. Titchener was ready to refute Dr. Meyers on behalf of the University of Cincinnati's medical team. He too asked the question, "Were old neurotic symptoms and problems laid bare by the trauma?" His answer differed from that of Dr. Meyers. "If this were so, we should have encountered a wide range of psycho-neurotic reactions. Although there were differences in modes of response, the uniformity of the syndrome in the Buffalo Creek survivors was striking." Dr. Titchener would show that the symptoms shown by the Buffalo Creek survivors were very uniform, irrespective of the mental make-up of the plaintiffs before the disaster.

As a layman, I dealt with Dr. Meyers's "Catch-22," "preexisting vulnerability" in a less scientific way. I assumed there is a point beyond which none of us can be pushed before we finally crack. Had this disaster pushed merely a few people over the brink, then it could have been argued that those people had a preexisting vulnerability or mental condition for which Pittston should not be held responsible. But this disaster was so overwhelming that almost everyone suffered mentally. Thus, it shouldn't matter what the mental make-up of the plaintiffs was before the disaster.

Dr. Titchener also asked and answered the obvious question about the psychiatric phenomenon they discovered—"Is it in individuals with weak egos exaggerating their complaints for the sake of winning a lawsuit? We think not, they did not make much of their complaints: they denied them."

We were also ready with Dr. Kai Erikson's study on the sociological aspects of this disaster. When we first contacted Dr. Lifton, he suggested we add Dr. Erikson to our team. He is chairman of Yale University's American Studies Department. As a sociologist, he was able to distinguish two facets of the psychic impairment prevalent in the survivors of the Buffalo Creek disaster—individual trauma and collective trauma.

Dr. Erikson identified the individual trauma as a "blow to the psyche that breaks through one's defenses so suddenly and with such force that one cannot respond effectively to it—they suffered deep shock as a result of their exposure to so much death and destruction, and they withdrew into themselves feeling numbed, afraid, vulnerable, and very alone."

Collective trauma, which is the more sociological type of trauma suffered by the Buffalo Creek disaster victims, is a "blow to the tissues of social life that damages the bonds linking people to each other and impairs the prevailing sense of communality." This collective trauma is a more gradual shock than the individual trauma of the disaster.

He pointed out that either of these traumas could take place in the absence of the other. A person could suffer deep psychic wounds as a result of an automobile accident but never lose contact with the community in which he is a part. Thus he would suffer only individual trauma. Similarly, those who have been uprooted from their community in slum-clearance projects can be said to suffer collective trauma, from the withering away of the surrounding community which supported them, although they do not suffer individual psychiatric trauma. In the Buffalo Creek disaster, the two traumas occurred together. Both were part of the psychic impairment suffered by each disaster victim.

This is illustrated by a young miner who left his home two hours before the dam failed. He was not in the flood waters, although he did view the aftermath of the flood. The miner lost many relatives in the disaster, but he did not evidence any strong symptoms of individual trauma immediately after the disaster. In fact, he did not begin to show any serious survivor symptoms until months later when he found a place to move back in the Valley.

Only then did he realize that his life had disintegrated. Pittston's doctors recognized this in him over a year and a half after the disaster:

> He is troubled by memories of the flood disaster. He feels some guilt that *maybe* he could have done more to help others prior to the disaster. He lost many relatives in the flood and has felt much prolonged grief over them. His

> family was almost like a clan or tribe to him, they being his chief source of friends and companions in almost all aspects of his life. He was dependent upon these relationships and now feels at a loss without them. I believe he was secure in the pre-flood environment of his family, because he was protected and his needs were met by a nurturing family. Now, forced to meet his needs on his own initiative and resourcefulness, his basic inadequacies are forcing the appearance of overt symptoms. He could well deteriorate.

But as Pittston's Dr. Meyers emphasized, "the time factors of onset of several of the troublesome symptoms . . . reveal that many months passed between the time of the flood and the onset of the new symptoms." Thus, Dr. Meyers concluded that Pittston was not responsible for this man's troubles.

Dr. Erikson would respond to this. He had spent days and days interviewing the people in the Valley, plaintiffs and nonplaintiffs. He got to know them as well as any of us. He found that "many of the traumatic symptoms experienced by the people of Buffalo Creek are as much a reaction to the shock of being separated from a meaningful community base as it is a reaction to the shock of being exposed to the actual disaster itself." Pittston's destruction of communities was most significant because, as he pointed out, " 'community' means a good deal more in Buffalo Creek than it does in most other parts of the United States—in Buffalo Creek, tightly knit communal groups were considered the natural order of things, the envelope in which people lived."

I was persuaded by Dr. Lifton and Dr. Erikson, but I had misgivings about the reception they might receive from a West Virginia jury. I had to find an expert in the state who could make these psychic-impairment claims more understandable and more credible for West Virginians.

Much to my surprise, I discovered, only a few months before the trial date, that there was a psychologist in

Charleston who had seen even more survivors than Dr. Meyers, and he'd also seen them over a longer period of time, in the Valley. Dr. Robert Kerns, the associate supervisor of professional services for the West Virginia Department of Mental Health in Charleston, had been traveling to Logan every Friday for almost two years, supervising the efforts of the Logan-Mingo County Mental Health Clinic to provide mental-health care for the disaster victims. From May 1972 through December 1972, he also went into the Buffalo Creek Valley, every week, with mental-health workers, knocking on trailer doors, seeking people who needed help but were afraid to go to the clinic.

Dr. Kerns told me the survivors suffered a "very specific disaster syndrome which they displayed fairly uniformly. I never made a visit to a trailer where I did not see the same pattern. I didn't go anywhere where I did not see quite severe anxiety reactions—loss of appetite, sleeplessness, extreme fear, and anxiety anytime it rains."

If I had tried to create an expert witness I couldn't have done better. He could make the jury of his fellow Charlestonians understand the seriousness of our survivor-syndrome claims.

Dr. Kerns had also been supervising the work of a number of pastoral counselors who had been providing lay counseling for the survivors. The Church of the Brethren, the Presbyterian Church U.S., and the United Methodist Church had answered a call for help from Reverend R. Judson Alford II, the chaplain at the Man Appalachian Regional Hospital. By June 1972, a three-pastor team had moved to the Valley. They were Reverend Robert Newcomb, a Presbyterian U.S. pastor who was completing his second-year chaplaincy at the Appalachian Regional Hospital's chaplaincy and pastoral-care education program at Beckley, West Virginia; Reverend Glen Sage, a Church of the Brethren pastor who was a first-year student in the program; and Reverend Vaughn Michael, a United Methodist pastor and a certified supervisor in the Association of Clinical Pastoral Education. Vaughn Michael stayed almost six months. Bob Newcomb stayed about fifteen months. And Glen Sage stayed twelve months, going home on weekends to Beckley. They operated much like psychiatrists, only their training was less medical, more directed toward grief therapy.

They confirmed the widespread mental suffering in the Valley. Indeed, each had written articles about his findings and some of his cases. If any jurors turned out to be devout Presbyterians, Methodists, or Church of the Brethren—look out, Pittston.

XXVI

A Settlement Meeting

Zane finally called back to set up a settlement meeting for March 28 at our old stomping grounds, the Holiday Inn in Charleston.

Bud Shay and I prepared our strategy. We agreed that we had to make sure that Pittston could not come away from this meeting with any excuse to go to Judge Hall and claim that we were being intransigent. We had to make it appear that we were interested in settlement and would be flexible. I speculated that Pittston would offer us something like $2.8 million. If they did, I thought we should tell them they still did not understand our case, otherwise they would not offer us only $2.8 million. But I would not be critical of their lack of understanding. I would say it was partly our fault, since we had not yet turned over to them our medical reports from the University of Cincinnati doctors.

Bud and I attended the meeting for our side. Zane was accompanied by W. Lauck Walton and Dan Murdock of Pittston's Donovan Leisure law firm. Lauck Walton is senior to Dan Murdock in this firm, but he seldom appeared in our case. I didn't know what to make of his presence at the meeting.

Zane opened the meeting. "As I told you on the phone, we do not intend to give you $32.5 million. We really would like to know what your real figure is. How did you get the $32.5 million?"

I tried to evade his question. Zane finally saved me, after I stammered around for a while. He said, "We have carefully studied each of these cases. A number of them are incredible. You are suing for a girl who died many months after the disaster in a fire. Some of the people that you claim psychic impairment for were not even in the Valley when the flood occurred. Some were not there till much later."

We jockeyed back and forth for quite a while. Finally Zane said he was ready to give us their settlement figure. But "the press has been too interested in this case and we want to keep this figure to the lawyers."

I said, "We will have to tell the plaintiffs what the proposal is, unless it is not one worth passing on."

Zane asked Bud and me to step outside so he could talk with Pittston's other lawyers. Out in the hallway we tried to guess what would happen. I surmised that Pittston would offer $3 million.

Zane called us back in the room. Before he could begin, though, Lauck Walton joked, "Where do you want the money delivered?" His remark seemed to embarrass the others in the room. It clearly embarrassed me that he was treating the matter so lightly. Zane then said, "Since you and Bud have agreed not to report our own settlement proposal to your clients if you deem it insufficient, we are ready to state our proposal to you. I am prepared to recommend to Pittston that they settle all these cases for a total of $3 million."

I asked, "Is your total figure broken down by individual plaintiff?"

He replied, "Arnold & Porter knows more about each

plaintiff than we do. You can break the figure down yourselves. I would be creating a problem for you if I offered one figure for one plaintiff and you want a different figure for that particular plaintiff. For example, I assume Arnold & Porter might want to pay Mr. Cowan more than Pittston would like to pay him."

Before I could turn down his $3 million offer, though, he quickly said that he would like to meet with us again in ten days in Washington. He said he would have to be in Washington to visit an aunt in a rest home there and could meet with us at that time. He appeared to be rushing on to propose this second meeting, figuring that we would not accept his $3 million offer, but hoping to keep the settlement talks going.

I went into my act. "I'm not surprised at the figure you have offered. Indeed, I guessed the exact dollar amount in the hallway before we came back in. I had assumed that serious settlement talk is still premature, and your proposal confirms my assumption. You will need to see our doctors' reports and learn more of the facts of reckless disregard before you will be in a position to make a reasonable settlement proposal."

They reacted violently. They thought I was accusing them of ignorance. They said they were in no way ignorant about this case.

"I'm not accusing you of ignorance. There just are a number of facts that I have not told you or made clear to you. In particular, your constant reiteration of the corporate-veil issue indicates to me that you do not understand all the facts of Pittston's direct involvement in this disaster."

Zane was not persuaded.

I deliberately left it up in the air as to whether we would talk again, or when.

XXVII

"Absent Plaintiffs"

On April 1, three days after our settlement meeting, Pittston hand-delivered to me a motion to dismiss the personal-injury claims of thirty-three plaintiffs, referred to by Pittston as "Absent Plaintiffs."

Pittston pointed out that, at the time of the disaster, these thirty-three absent plaintiffs "were in jail or were in hospitals in other parts of the state. Still others were out of the state in Florida, Kentucky, Ohio, Maryland and even New Mexico." Pittston added, since "there is no genuine issue as to the material facts that none of the 33 plaintiffs was physically injured in the course of the flood on Buffalo Creek on February 26, 1972, that none was threatened with immediate physical injury by the flood, and that none witnessed any other person sustaining a physical injury in the flood. . . ," the psychic impairment claims for these thirty-three plaintiffs must be dismissed by the court as a matter of law.

Pittston had struck at our Achilles heel by moving to

dismiss the psychic-impairment claims of only these thirty-three "absent plaintiffs" rather than the psychic-impairment claims of all plaintiffs not touched by the water that day. From the outset, we recognized that it would be almost impossible, as a legal matter, to sustain the psychic-impairment claims for these particular thirty-three plaintiffs. Judge Hall's early rulings in the case had highlighted this problem. When Judge Hall said he wanted to try a few representative cases, and described the possible representative categories, he did not even include the possibility of recovery for someone who was not in the water and who did not see anyone else injured in the water, and who, indeed, was absent from the Valley on the day of the disaster. This posed a difficult dilemma for us. Should we agree with Pittston that these thirty-three cases should be dismissed?

I was afraid that if we did not dismiss these thirty-three "absent plaintiff" cases, Judge Hall might rule, as a matter of law, that anyone who was not in the water and did not witness others being injured had no right to collect for psychic impairment. That kind of ruling, though addressed only to the thirty-three so-called absent plaintiffs, could undermine the psychic-impairment claims of almost all the other plaintiffs, since many of them had gotten up the hill before the water came through, did not look back, and did not witness anyone in the water. This was especially true of many of the children whose parents rushed them out of the Valley and shielded them from any view of the immediate disaster. On the other hand, if we stipulated with Pittston that these thirty-three cases could be dismissed, maybe we could hold our losses to just these thirty-three.

But I doubted that Pittston would be satisfied with a dismissal of only these thirty-three psychic-impairment claims. Why did Pittston hand-deliver its motion to me on April 1, rather than mail it? When you receive a motion in the mail, you have ten days from the date of mailing, plus another three days because of the mails, to respond. But when a motion is given to you in person, you have only the ten days in which to respond. When Pittston handed me its motion on April 1, they said that Judge Hall already had set it for a hearing on April 11 at two-thirty in the afternoon, exactly ten days later.

We were surprised that Pittston had arranged a hearing date without checking with us. We called the court and learned that Pittston had also scheduled two additional hearings with Judge Hall on April 24, one at nine-thirty in the morning, the other at one-thirty in the afternoon. Pittston had not filed any motions in connection with these hearing dates. But we could guess what they'd be. Apparently Pittston figured that Judge Hall would readily agree to dismiss the psychic-impairment claims of the thirty-three "absent plaintiffs." Then, relying upon Judge Hall's ruling as to these thirty-three plaintiffs, Pittston would file its next motion to dismiss another category of psychic-impairment claims—maybe those who ran up the hill to safety and were not touched by the water. Pittston could hand-deliver this next motion on April 12 or April 13, giving us ten days' notice before the hearing already blocked out for April 24.

They probably assumed that on April 24 they would get another ruling from Judge Hall dismissing some more psychic-impairment claims. Then they would be ready to file their final blockbuster motion, prior to the May 1 deadline for the filing of all motions. Relying on two favorable rulings from Judge Hall, they then would move to dismiss any remaining psychic-impairment claims not connected with some contemporaneous physical injury. There wouldn't be much left to our case after that.

Pittston was trying to slice up our case bit by bit. We stopped these "salami tactics" by obtaining an extension of time to respond to their "absent plaintiffs" motion, a motion Bud Shay always referred to as the "absent employer's motion." This forestalled any effort by Pittston to file its motions, piece by piece, as to the personal-injury claims of any of the plaintiffs. And it gave us the necessary time to file a carefully prepared brief with Judge Hall on this critical legal issue.

A federal judge is required to decide cases on the basis of the substantive law of the state in which he is sitting. Thus, Judge Hall would have to decide the rights of absent plaintiffs in the same way that a West Virginia state court would decide the case.

The few decisions of the West Virginia Supreme Court on recovery for mental injury provide a capsule glimpse of the development of the law on this subject throughout the United States. In West Virginia, as elsewhere in this country, the courts have moved slowly, much more slowly than juries, in permitting any recovery for mental injury.

Thus, in 1899 the West Virginia Supreme Court overturned a jury verdict of $500 for Claude Davis for mental suffering unconnected with any physical injury. The Western Union Telegraph Company had negligently delayed delivery of a message to Claude Davis to "Come at once," so he missed his mother's funeral. He sued the telegraph company for $1,500 damages. The jury agreed that the telegraph company's negligence had caused him mental suffering and awarded him $500. The West Virginia Supreme Court, though, reversed the jury. The court, quoting from an English law text, explained the law's unwillingness to permit juries to award damages for mental suffering alone:

> Mental suffering alone, and unaccompanied by other injury, cannot sustain an action for damages, or be considered as an element of damages. *Anxiety of mind and mental torture are too refined and too vague in their nature to be the subject of pecuniary compensation in damages*, except where, as in the case of personal injury, they are so inseparably connected with the physical pain that they cannot be distinguished from it, and are therefore considered a part of it. [Emphasis added.]

So Claude Davis lost his $500. His "anxiety of mind" was "too refined and too vague" for the law at the turn of the century.

Almost fifty years later, in 1945, the West Virginia Supreme Court spoke again on the subject of mental suffering. Mrs. Theresa Monteleone was riding in her son's Studebaker car in Wheeling, West Virginia, when a broken electric trolley wire shattered the car's windshield and showered her with small slivers of glass. She suffered only a slight cut on her face. However, two weeks later she began to suffer severe psychological effects that required medical attention. She sued the Co-Operative Transit Company for her mental and emotional disturbances. Her psychiatrist testified that she had suffered "a post-traumatic psychoneurosis." The jury agreed, and awarded her $5,000. Again the Supreme Court of West Virginia disagreed, and reversed the jury's award.

But the *Monteleone* judges in 1945 did not revert back to the dark ages of 1899 and the *Davis* case. They expressed much greater faith in modern psychiatry. They explained that "the Davis case was decided before the theory that a person has 'a legal right to mental tranquility' had been developed to the extent that it now has." But they still didn't feel it had been developed enough to let Mrs. Monteleone keep the $5,000 the jury awarded her.

The court did agree that there were "three fairly well settled principles allowing recovery" for "mental and emotional disturbances caused by the wrong of another." First, "mental disturbances that accompany or follow an actual physical injury caused by impact." This was what Mrs. Monteleone claimed. But the court said there was only a slight impact and there was no relationship between the "cut upon her face about the size of a pimple" and her severe mental injury. Similarly, if the plaintiffs in our case were seeking recovery for mental suffering based upon some mere touching of their bodies by the black water, the relationship between their serious nervous and emotional disturbances and that slight impact would be too tenuous.

But our plaintiffs could try and fit into the second and third categories of permitted recovery referred to by the *Monteleone* judges. Second, "no impact and no physical injury at the time, but a physical injury afterwards results as the causal effect of a nervous shock." Maybe we could convince Judge Hall that a nervous disorder is itself a physical injury. A federal judge in Virginia so held in 1960:

> It is unreal to attempt to distinguish between mental and physical injury. An affront to either the mental or the physical sensibilities is an affront to the personal being. The only question in Virginia is whether the "damage" is of substance and sufficiently identifiable in the person of the claimant.
>
> Hence, we need not stop to ponder here whether injury to the body followed injury to the emotions or the reverse was true. . . . [T]he inquiry is simply whether there was a personal injury, either mental or physical.

Thus, the survivors might recover because of Pittston's negligent conduct, if Judge Hall decided to throw out the ancient mental-physical distinction.

Monteleone's third category might also be applicable—"no impact and no physical injury . . . but an emotional and mental disturbance . . . the result of the defendant's intentional or wanton wrongful act." In other words, the plaintiffs might recover for their mental suffering, even if it is not considered physical, if we could prove Pittston's conduct was "intentional or wanton," rather than merely negligent.

But what if Pittston was only negligent, and mental injury is distinguishable from physical injury? The *Monteleone* court talked about that in 1945—"recovery for mental and emotional injury alone, caused by the defendant's simple negligence." The West Virginia Supreme Court wasn't ready to permit recovery under those circumstances in 1945.

Pittston read the *Monteleone* case as supporting their motion to dismiss the absent plaintiffs' claims. "Since West Virginia under the Monteleone rule denies a right of recovery to a plaintiff who fails to suffer physical impact caused directly by the tortious conduct alleged, it follows *a fortiori*, that it denies a right of recovery for personal injury to plaintiffs who, like the Absent Plaintiffs, unquestionably were not within the zone of physical danger." This "zone of danger" test was another exception some courts had carved out to permit recovery where there was no physical impact. For them it was enough to have been in the zone of danger. But West Virginia never adopted that exception, and our absent plaintiffs couldn't meet it anyway.

Pittston's lawyers also relied on the old 1899 *Davis* case. They cited a recent opinion by Judge Christie which indicated his belief that *Davis* was still the law in West Virginia. Judge Christie had written, in 1967, that "under West Virginia law, one may not have a recovery for mental suffering alone, and unaccompanied by any physical injury." He cited the 1899 *Davis* case for this rule of law. Luckily, we weren't before Judge Christie. But Judge Hall would not easily disagree with an interpretation of West Virginia state law by this respected West Virginia federal judge.

We replied with our interpretation of *Monteleone*, emphasizing the *Monteleone* court's feeling that *Davis* was no longer good law. More important, we told the Court we believed the *Monteleone* case in 1945 placed West Virginia as "one of the most progressive states in permitting recovery for mental suffering," because the court there said there could be recovery for mental injury alone, if the defendant's conduct was wanton. We argued that a jury could decide that Pittston's conduct was wanton.

After filing our memorandum, we appeared before Judge Hall to argue Pittston's motion. Judge Hall gave no indication of his feelings on the matter, only informing us that he would issue a written opinion sometime in the future. This was the first time he had not ruled at the completion of the oral arguments. That was good news for us. Judge Hall was carefully considering Pittston's attempt to dismiss the claims of these thirty-three plaintiffs.

There was more good news the next day. The president of the West Virginia State Bar Association reported to the press that after "a thorough and protracted investigation" its investigators "did not uncover any concrete evidence to support a charge against anyone for engaging in the unlawful practice of law" in connection with the Buffalo Creek disaster. Hardly ringing praise for our efforts, but at least they had finally finished with us.

XXVIII

"Unavailable" Witnesses

As we neared the deadline set by Judge Hall for the completion of discovery, we lost two friendly witnesses we thought we could count on to testify for us at the trial.

We hoped Robert O. Weedfall, West Virginia's state climatologist at the time of the Buffalo Creek disaster, would help us because of the statements he made about the weather conditions at the time of the disaster. The *Morgantown Dominion-News* had quoted him as follows:

> State Climatologist Robert O. Weedfall said last night persons who blame the Logan County flood on weather or topography are "silly" and "asinine."
>
> In discussing the tragedy last weekend which claimed at least 84 lives, Weedfall said the expression "Act of God" has been widely quoted.
>
> "That's a legal term," he said, "but there are other legal terms like involuntary manslaughter because of stupidity and criminal negligence."

> Weedfall maintained if the flood had not occurred this year, "it would have come next year or the next year, but it had to come."
>
> A potential hazard exists when any dam is constructed of slag, he said, because the material is constantly burning and becomes hollow. "It's going to burn out—it's like a time bomb," he maintained.
>
> * * *
>
> Weedfall said the 3.72 inches which fell on Buffalo Creek hollow from February 24–26 is not uncommon to the area.

We had written Mr. Weedfall to ask him to testify for the plaintiffs. He called us back, collect. He was willing to testify for us for a fee of $500, with $100 down and $400 additional for the testimony. We agreed. But later he indicated his desire for an additional $10 per plaintiff if, and only if, we were succesful in our lawsuit. With over 600 plaintiffs, this meant a fee of $6,000 instead of $500. He'd upped the ante quite a bit.

We wrote him that we could not agree to pay him any additional money if we were successful in our suit. "It would not be appropriate for us to agree to a contingent fee arrangement for an expert witness, since this would allow counsel for Pittston to suggest that your testimony was based on hope of financial gain rather than an expert analysis of the available facts. Although this would not be true, it might have an adverse impact on the jury."

We soon had Mr. Weedfall's response. He informed us that from now on, if we had any questions for him, we should contact Zane Grey Staker.

I doubted that Zane Grey Staker had any intention of having him testify at the trial as Pittston's expert. Mr. Weedfall could hardly be expected to help Pittston's case, since I would have the opportunity to cross-examine him and bring out the statements he had made concerning manslaughter and murder. Indeed, his testimony would be even more damaging to Pittston if he testified as its expert.

I was afraid, though, that Mr. Weedfall might not be available at the trial, even as a Pittston expert. He might find it convenient to take a vacation during the trial, outside the subpoena power of the court, so I subpoenaed him for an immediate deposition. I then could read his deposition testimony under oath to the jury at the trial if he decided to absent himself from the jurisdiction at the time of the trial.

Pittston responded that it was improper for us to depose their expert. Judge Hall resolved the matter by informing Pittston that they had the responsibility to make certain that Mr. Weedfall would be available to testify at the trial, if the plaintiffs wanted his testimony.

We were also having problems with William Wahler, the engineering consultant from California whose firm prepared a comprehensive, million-dollar study of the Buffalo Creek disaster for the Bureau of Mines. Mr. Wahler had agreed to come from California to Washington to discuss his study with us. However, just before the meeting he suddenly notified us, "I do not believe my testimony would be useful to your case. I believe, therefore, that you should find someone else more available to meet your needs."

It was hard to imagine how his testimony could not be helpful to us, since his study demonstrated, in conclusion after conclusion, that the dams on Middle Fork were recklessly built and maintained. When we pressed him, we learned the real reason why his testimony "would not be useful to our case." This time he sent a telegram informing us that he had been retained as an expert by the Pittston Company. Again, as in the case of Mr. Weedfall, I did not believe that Pittston intended to put Mr. Wahler on the stand as its expert in this case. So, as with Mr. Weedfall, we had to find some way to get this Pittston expert's testimony before the jury.

Since Mr. Wahler, unlike Mr. Weedfall, lived outside the

subpoena power of Judge Hall's West Virginia federal court, we couldn't get Judge Hall to order Mr. Wahler to be available for us at trial, as he had with Mr. Weedfall. However, a California federal district judge has the authority to order Mr. Wahler, and other Californians, to appear, in that state, for a deposition before trial. This deposition could be read to the jury at our trial. So we asked the California federal court to order Mr. Wahler to appear for a deposition.

Mr. Wahler fought our deposition request tooth and nail. He hired his own counsel, and Pittston's counsel also joined in, in a strenuous effort to keep us from taking his testimony. But the judge in San Francisco ordered Mr. Wahler to appear for our deposition in California to testify about his government study.

At his deposition we had Mr. Wahler repeat, under oath, the conclusions he had set forth in his report. Now we could read that testimony to the jury as Pittston's expert's conclusions, to add to those of our own experts, Mr. Davies and Mr. Fuquay. And we could insist that Mr. Weedfall testify at the trial, if Pittston continued to push its argument that "excessive rainfall" caused the Buffalo Creek disaster.

XXIX

The "Glaring Deductible"

By this time our continuing efforts to obtain information on Pittston's insurance coverage began to provide additional evidence of Pittston's recklessness. At the outset of the case, I had requested that Pittston produce all documents, for a specified time period, from the files of W. J. Kelleher, Pittston's insurance manager. Pittston had refused, so I deposed Mr. Kelleher with only a few insurance documents Pittston had turned over in response to some of our other document requests.

I asked Mr. Kelleher about the postdisaster insurance agreement, which referred to "certain differences" with respect to Pittston's insurance coverage. He responded that he was the one who recommended that Pittston increase its insurance coverage before the disaster, but his recommendation was not related in any way to the Buffalo Creek disaster. He had made it long before the disaster, and it was just coincidental that the increased insurance coverage was finalized on February 24, 1972, two days before the disaster.

On the other hand, Mr. Kelleher's testimony on a "$400,000 insurance gap" for dam damage, referred to in one of the Pittston documents, was helpful to our case. Mr. Kelleher remembered that when Pittston's insurance coverage came up for renewal at the end of 1970, Lloyd's of London, one of the insuring underwriters, asked for information on all Pittston's dams. Mr. Kelleher assumed this request was not prompted by any specific problem the insurance underwriters had with any Pittston dam. "We did not attach much importance to the dam exposure. We had no reason to. We had never received any complaints from the people regarding dams and so forth. We thought that maybe Lloyd's was considering other risks where they had problems and we didn't believe it belonged to us."

Mr. Kelleher got together some information on some of Pittston's dams and forwarded this to Albert P. Bedell of Johnson & Higgins, the independent insurance broker used by Pittston. Mr. Kelleher recalled sending some photos and information on a Pittston dam at Dola, West Virginia.

After the insurance underwriters received this sketchy information on Pittston's dams, they added a $1 million deductible on any property damage caused by any Pittston dam failure. Mr. Kelleher was shocked at this "glaring deductible," as he called it. He immediately attempted to "get that [$1 million deductible] out of there." Through Johnson & Higgins, Pittston was able to persuade Lloyd's of London and the other insurance underwriters to reduce this deductible to $500,000. Pittston and Johnson & Higgins continued their efforts to "delete" even this $500,000 deductible for dam damage. They assured the London underwriters that Pittston had only one "major dam," the dam at Dola, West Virginia. But the underwriters would not reduce the deductible any further. Thus, at the time of the Buffalo Creek disaster, there was a $500,000 deductible on dam damage. Pittston's primary insurance coverage in this country would pay the first $100,000 of this uninsured

$500,000, leaving an insurance gap on dam damage of $400,000.

All this was news to me. I immediately asked Pittston to search its files for all the documents referred to by Mr. Kelleher in connection with the renewal of Pittston's insurance at the end of 1970. In particular, I wanted copies of all the documents on Pittston's dams which Pittston collected and forwarded to its insurance broker, Johnson & Higgins, and I wanted information on Pittston's one "major dam" at Dola.

I was in for quite a shock. The documents, when produced, indicated that the Dola dam had failed in March of 1967, and Pittston's insurance company had settled claims made by a number of people whose property was damaged. One such person, Rex McGinnis, had sued Pittston before settling with the insurance company. I promptly asked Pittston for the documents on this lawsuit and telephoned Rex McGinnis and his lawyer.

I learned that on March 11, 1967, five months after the Aberfan disaster, and at almost the same time that Dam 1 overtopped on Middle Fork, a refuse-pile dam used to impound black plant water at Pittston's Clinchfield Coal Company mine at Dola, West Virginia, gave way after heavy rains. This unleashed a twelve-foot-high wall of black water on the valley below the dam. No one was injured, but one house was filled with coal refuse and a substantial amount of land was damaged.

Rex McGinnis sued Pittston in a state court, the Circuit Court of Harrison County, West Virginia. The legal papers filed in this lawsuit told a story now all too familiar. Pittston admitted that its "dam impounded a quantity of water and other materials" and that its "dam overflowed" on or about March 11, 1967. But Pittston denied all liability "by reason of an act of God." Nevertheless, Pittston eventually settled with Rex McGinnis.

It appeared from the date of this settlement, May 18,

1970, and the request by Pittston's insurance underwriters at the end of 1970 for information on all Pittston's dams, that there was a connection between the $1 million deductible for dam damage and this Pittston dam failure, despite Mr. Kelleher's belief to the contrary. However, my new request for documents from Mr. Kelleher's files on this renewal did not produce much. Although Mr. Kelleher had testified "Yes" at his deposition when I asked if he had a copy of the written proposal "that had this $1 million provision in it," his "yes" answer was changed to "no" after the deposition.

So I telephoned Mr. Bedell of Johnson & Higgins and asked him to check his files for me. Pittston is an important customer of Johnson & Higgins. So, when Johnson & Higgins said they could not help me, I served Mr. Bedell with a subpoena for his files and told him to appear for a deposition in New York.

Johnson & Higgins hired a lawyer, got an affidavit from one of Pittston's lawyers, and went before a federal judge in New York City to quash our subpoena and our deposition. They argued that the court should forbid "all inquiries and discovery which in any way concern any issues between the insured and insurers concerning insurance and relating to [Pittston's] application for insurance coverage." They said that information was completely irrelevant to our lawsuit.

The argument on Johnson & Higgins' effort to stop our deposition and document request was heard by Federal Judge Constance Baker Motley in New York City. This was a nice twist of fate. Judge Motley and I had been fighting for the same cause back in the civil rights days when she was an NAACP trial attorney. She had appeared often in the South on school desegregation cases at the same time that I was working for the Civil Rights Division on voting cases. I vividly recall seeing and hearing her argue before Federal District Judge William Harold Cox in Jackson, Mississippi. Her arguments fell on deaf ears. Judge Cox was not prone to rule in favor of anyone voicing a civil rights complaint. He didn't discriminate, however. He denied such claims whether they were argued by black lawyers for the NAACP or white lawyers like me for the United States government.

I hoped Judge Motley would recognize that the people we represented in West Virginia needed just as much judicial help today as did the blacks in the South during the civil rights days.

Brad Butler argued our motion to compel Mr. Bedell to appear, with documents, pursuant to our deposition request. As soon as Brad thought Judge Motley was about to rule for him, Johnson & Higgins's lawyer would stand up and begin arguing all over again. Then when Brad thought Judge Motley might change her mind and rule for Johnson & Higgins, he would jump to his feet and begin arguing his side all over again. Finally, Judge Motley ordered Mr. Bedell to appear for his deposition and to produce the documents we wanted.

My suspicions were correct. The request in 1970 from the London insurance underwriters for information on all Pittston's dams was tied directly to Pittston's 1967 Dola, West Virginia, dam failure. Travelers Insurance Company, Pittston's primary insurance carrier in this country, had settled Rex McGinnis's lawsuit in May 1970 for $28,401 plus $2,506 in expenses. Thus, since the insurance renewal form from London required that Pittston list all payments over $25,000 under the old policy, Mr. Bedell informed London of the McGinnis payment, explaining that a "Dam at Clinchfield Coal Company broke allowing mine refuse to flow over claimant's property causing damage to land, houses, garages, etc." This is the reason the London insurance broker then asked Mr. Bedell for "details" on all of Pittston's "dams." And even though Mr. Bedell and Mr. Kelleher responded with information on only three Pittston dams, the London underwriters then included the "glaring"

$1 million deductible in Pittston's new 1971 policy.

Although Mr. Bedell and Mr. Kelleher were able to reduce this "glaring deductible" to $500,000, when they pushed further, with numerous entreaties, to get it down to $100,000, the London broker telexed back—"[underwriter] adamantly declining agree reduce $500,000." The London brokers added that the lead insurance underwriter was "unprepared commit himself without an independent engineering report/opinion and warns that even if same apparently satisfactory he may or may not quote" a bid to cover the remaining $400,000 insurance gap on dam damage.

Mr. Bedell wrote Mr. Kelleher about this and said "the estimated cost would be approximately $5,000" for "an independent engineering report/opinion" on Pittston's dams.

Did Pittston pay the $5,000 to have its dams checked? It was time to depose Mr. Kelleher again and find out. But this time I'd get the remaining documents in his file first. I went back to Judge Hall. We'd come a long way from our original request for Pittston's insurance contract. Now I wanted the rest of the insurance documents to close the circle on the insurance question.

These documents, though they related to insurance coverage, also related to recklessness. Pittston's own insurance carrier had warned Pittston in a way every juror would understand. When we get two speeding tickets, our car insurance rates go up, because the insurance company's computer says that the odds for an accident increase when a driver has two speeding tickets, even if that person has never had an accident. When we get our rates increased, or our deductible increased, we know we'd better be careful, or next time we may lose our coverage altogether.

We told Judge Hall that Pittston knew, or should have known, to check the safety of its dams when it got hit with a million-dollar "glaring deductible" for dam damage, or at least when the lead underwriter "adamantly declined" to reduce the $500,000 deductible. Certainly they had been warned when their lead underwriter insisted on an "independent engineering report/opinion" on all Pittston's dams. We wanted the documents from Pittston's own files to prove all this.

Judge Hall ordered Pittston to produce them. Pittston did, unwillingly. "In making this production we don't waive our continuing objection to documents relating to the procurement of insurance coverage for claims that are the subject matter of this litigation."

When the documents were produced I had the final proof. There was no "independent engineering report/opinion" among them. Pittston hadn't bothered to obtain such a report, not for the $5,000 Mr. Bedell thought it would cost, and not even for the $500 Mr. Kelleher estimated it would cost.

XXX

"The Dam Was All Right"

Preparing a trial is like writing a stage play. How will it open? Who will testify first? One of our engineering experts, or someone from the Valley? Perhaps the first witness should just set the scene. Identify some aerial photos of the Valley so the jury can slowly begin to know what it looked like before the disaster? Or should I start with movies of the disaster, get their interest right from the beginning? But how can I sustain that interest and their concern? I want the jury angry at the end, just before they file out. Maybe I can let their attention wander a bit at first.

I had been thinking about this as I worked on the trial brief, the script for the trial. The trial brief summarizes the entire case and tells the judge what will happen at the trial. This permits the judge to study the legal questions before the trial begins, so he can quickly rule on them during the trial. Our trial brief also would list the witnesses in the order of their appearance. To prepare this list I had been constantly arranging and rearranging the characters in my head, telling them when to come on stage, trying to make it dramatic, to keep the jury alert.

But a trial differs in one major respect from a stage play. When a playwright writes a script, he knows the actors must follow the script. They are not supposed to ad-lib. But you can't trust the characters in the trial to follow your script. You can't completely control the show. The witnesses may make up their own lines on the stand, especially when Pittston's counsel would begin cross-examining them, putting new lines in their mouths.

But the similarities are still greater than the dissimilarities. At the end of our trial, the jury would vote on our drama much like the critics on the first night vote on the playwright's work. It would be instant life or death for us. The playwright may have another chance. He may open his play out of town, and if the critics hit home with their objections, he has a chance for a rewrite and for other opening nights before he hits the big time. But with our trial, there could be no opening in Philadelphia with a run-through a few times before we hit New York. We would open and close in Charleston only once. If our drama for our representative plaintiffs earned the plaudits of the jury, and a finding of reckless disregard and a large verdict, we would win. If not, we would lose, and all the other cases we would have to try thereafter also would lose.

Through the experts and some of Pittston's own people I thought we could show that Pittston's dam was recklessly built, contrary to simple engineering principles, and in violation of federal and state safety laws. We could also unfold the evidence of Pittston's disregard of the warnings from its own prior dam failures at Dola and at Lick Fork, the warnings from the failure of Dam 1 at Middle Fork, the warnings from the Aberfan disaster, and even the warnings from its own insurance underwriters. But I also wanted the jury to know about Pittston's reckless acts, and failures to act, during the last days and hours before Dam 3 collapsed. My problem was, how best to tell the story of this final, fatal week from all the facts I now had collected.

This story began on the Monday of the week before the disaster with Jack Kent, the superintendent of strip mining for the Buffalo Mining Company. Mr. Kent had no official responsibility for Dam 3. But he remembered the failure of Dam 1 in 1967, and he, like so many others in Saunders, was always afraid that Middle Fork's dams might fail again whenever it rained. So on Monday of this last week, he began looking at the dams on Middle Fork daily, "because there was a lot of water behind them and we were concerned."

By Tuesday, even Mr. Dasovich, the man in charge of the Buffalo Mining Company operation, was concerned by the rains. So he drove up to look at Dam 3, "to see how the water level was being affected."

Wednesday, as Mr. Dasovich later remembered it, there was "a lot of rain." More people began to be concerned about the dams on Middle Fork. According to Ben Tudor, Buffalo Mining Company's superintendent, "Actually, I would say the major concern began about the middle of the week."

On Thursday morning, Mr. Dasovich made his daily long-distance telephone call to Mr. Spotte, the president of the Pittston Coal Group, to report on the Buffalo Mining Company's coal production for the previous day. During this call, Mr. Dasovich also told Mr. Spotte that he was concerned that "the water was rising in No. 3 impoundment."

That afternoon Mr. Spotte, Mr. Kebblish, and Mr. Yates flew to the Buffalo Creek operation to review the strip-mine permit with Mr. Dasovich. None of them bothered to go and look at Dam 3, despite the fact that Mr. Dasovich had telephoned Mr. Spotte that morning of his concerns for Dam 3, and despite the fact that Mr. Kebblish testified that Mr. Spotte and Mr. Dasovich discussed the rising water behind the dam while they were driving in the car at Buffalo Creek that Thursday afternoon.

But Mr. Kent checked Dam 3 on Thursday afternoon after he got off work. There was no water gauge behind Dam 3, so he got a stick, about three feet nine inches long, and placed it at the edge of the water behind the dam. When he saw Mr. Dasovich at the office on Thursday evening, he told him he'd keep a check on the dam. Jack Kent madc his last check on Thursday at eleven-thirty that night.

On Friday morning before six, Mr. Dasovich went to check Dam 3. He either did not know about or did not remember Mr. Kent's stick, so he did not have any way of accurately determining if the water behind Dam 3 had risen. Without a water gauge, and unaware of Mr. Kent's stick, he still decided that the water "had receded a couple of inches." Mr. Dasovich then called Mr. Kent and told him not to go and check the dam any further. Mr. Dasovich did not go back to the dam again on Friday because he now assumed the water had gone down.

Mr. Kent kept checking Dam 3 despite Mr. Dasovich's call. After work on Friday, Mr. Kent's stick showed that Mr. Dasovich's guess was wrong. The water had not receded. Instead it had increased eighteen inches from Mr. Kent's last check on Thursday night and now was only a few feet from the top of the dam.

All day Friday, Mr. Kebblish, Mr. Spotte, Mr. Yates, and Mr. Dasovich reviewed the proposed strip-mine operation. They were only three to four miles from Dam 3. It would have taken only a "very short time" to drive over and let Pittston's chief engineer, Mr. Yates, make an inspection. They did not bother to do so. Instead, on Friday night, Mr. Kebblish, Mr. Spotte, and Mr. Yates all left the Buffalo Creek area. Mr. Kebblish and Mr. Spotte flew out by helicopter, Mr. Kebblish to the Charleston airport, and Mr. Spotte to Dante, Virginia. Neither bothered to have the helicopter fly over Dam 3.

On Friday night it began to rain heavily again. There were radio announcements of flash-flood warnings and forecasts of thunderstorms. Mr. Kent, on his own again, and contrary to Mr. Dasovich's statement that there was no need to check the dam, began checking Dam 3 every two hours, all Friday night and into the early hours of Saturday morning.

The people in Saunders also were worrying about Dam 3 that Friday. That afternoon, Mrs. Maxine Adkins had great trouble staying at work. There had been warnings about flooding along the Guyandotte River, the major river that Buffalo Creek empties into. The Guyandotte runs north from Man up to Logan, the county seat, and then beyond there. It floods often.

These flood warnings made Mrs. Adkins very nervous. She remembered the failure of Dam 1 in 1967 and always became frightened when it rained. When her boss asked her what was bothering her, she said she was afraid the dam would break, with her husband home alone without a car. Her boss told her she could leave early and go on home. She left immediately. While driving home, she heard on the car radio that anyone needing help because of any flooding from the Guyandotte River should call the National Guard at the jailhouse in Logan.

When Mrs. Adkins got home, she tried to convince her husband that they should leave Saunders. He had been in the hospital when Dam 1 failed in 1967 and did not share her strong fears that Dam 3 might now fail. He did not want to leave. However, a few hours later a neighbor told Mr. Adkins he thought the dam would not make it through the night. Her husband believed him.

So, late that Friday night, Mr. and Mrs. Adkins and some of their neighbors went to the schoolhouse at Lorado, about three miles down the Buffalo Creek Hollow from Saunders. They thought they'd be safe there. Mrs. Trailer opened the Lorado schoolhouse for them.

While in the schoolhouse, Mrs. Adkins started talking to Mrs. Trailer about her fears. She said she had heard on the radio that people should call the jailhouse if they needed help from the National Guard. Mrs. Trailer called Mr.

Ramey, the principal of the Lorado school, and got his permission to open his office so Mrs. Adkins could make a long-distance call to the jailhouse in Logan. Mrs. Adkins called the jailhouse about 3:30 Saturday morning.

She told the jailer she was afraid the Buffalo Creek dam would break. He said, "The Buffalo Creek dam?," apparently not knowing what that was. She told him about the dam up Buffalo Creek and that her neighbors thought it would break. The jailer took her name and said he would see what he could do. When he asked her phone number, she told him she was at the Lorado schoolhouse and gave him that number.

Sheriff Grimmett remembered receiving a phone call from the jailer sometime early Saturday morning, about 4. The jailer told him a lady had called from a schoolhouse in the Valley to warn him that a group of people had gathered in this schoolhouse because of their fears for the dam. The sheriff's wife then telephoned this lady at the schoolhouse.

Mrs. Grimmett asked what the problem was. Mrs. Adkins told her what her neighbors had said, that the water was about to get over the top of the dam and that the people should get out. She said older people who had lived there for years said the dam couldn't hold until daylight, and that is why she called. She said she wanted to have the National Guard come out and warn the people.

After talking to Mrs. Adkins, the sheriff's wife telephoned Harold Myers about 6:00 A.M. Harold Myers worked for the Buffalo Mining Company, and she thought he might know something about the dam. Mr. Myers said he would talk with someone at the Buffalo Mining Company and call the sheriff back.

Harold Myers immediately telephoned Ben Tudor, Buffalo Mining Company's superintendent. Mr. Myers told him the sheriff was thinking "about calling the National Guard" because some people were worried about Dam 3. Mr. Myers asked him "if there was any problem with the dam, and he said that they had been working on it and it was O.K. now." So Mr. Myers called the sheriff back between 6:30 and 7:00 A.M. and told him there was no trouble with the dam.

The chain was now complete. The people at the schoolhouse that night, frightened that a dam on Middle Fork would fail again, telephoned the jailer to call out the National Guard; the jailer telephoned the sheriff; the sheriff's wife telephoned the school, talked to Maxine Adkins, and learned of the people's fears; the sheriff's wife telephoned Harold Myers of Buffalo Mining Company; Harold Myers telephoned Ben Tudor; Ben Tudor—Buffalo Mining Company's superintendent—said everything was all right; and Harold Myers then telephoned the sheriff and told him this.

Meanwhile, Mr. Kent had been checking the dam all night. The water was now about a foot from the top, rising about three inches an hour. He realized Dam 3 soon would overtop and thought that the mobile home nearest to the dam in Saunders might be washed down Buffalo Creek Valley unless they did something fast. So he telephoned Mr. Dasovich, sometime between 5:00 and 5:15 A.M.—much earlier than he had ever called him before. He woke him and told him the water had been rising two inches an hour and now was rising three inches an hour. He told Mr. Dasovich he should come up to the dam.

At about this same time, the sheriff, prodded by his wife, decided to send two deputy sheriffs up to Buffalo Creek to check on the dam and see what all the worry was about.

Mr. Dasovich took his time getting up there. He first stopped at a coffee shop in Man to have his morning cup of coffee. There he ran into one of the deputy sheriffs who was on his way up to the dam. Mr. Dasovich followed him on up the creek. He arrived at the dam around 6:00 A.M. He had not been to check on the dam in twenty-four hours—since 6:00 A.M. on Friday morning—even though the water level had been rising rapidly behind Dam 3 during that time, and even though there had been major concerns about the dam at least since Wednesday. Now, on Saturday, at least a year after Dam 3 began to impound Middle Fork, at least ten months after the West Virginia Department of Natural Resources had said that Dam 3 needed an emergency spillway or overflow, Mr. Dasovich decided that something should be done about Dam 3. He told Mr. Kent to put in two twenty-four-inch pipes to drain the water behind Dam 3 into the ditch beside the dam. But it was much too late for that. The water was almost at the top of the dam, and, apparently unknown to Mr. Dasovich, the dam was about to break.

Mr. Dasovich made no effort to get out a general alarm, or any alarm, to warn the people. Instead he told the deputy sheriff, and the frightened people huddled together near the Lorado schoolhouse, that they didn't "have anything to be concerned about." Regretfully, they accepted his word and said, "That's good enough for me."

Mr. Dasovich then called long distance to Mr. Spotte to give his morning report, but this time he called him an hour earlier than normal—somewhere between 6:00 and 6:30 A.M. He told him "the water behind the No. 3 impoundment had gone up again"; that "somebody had called the sheriff's department and notified them of the situation there, that it was becoming hazardous, dangerous"; and that "some of the local people in the town of Saunders were very much concerned."

Did Mr. Spotte make any suggestions? No. Did Mr. Spotte say that maybe he would send Mr. Yates, Pittston's chief engineer, to check on this hazardous situation? No. Mr. Spotte says he just left the problem to Mr. Dasovich. Was Mr. Dasovich qualified to determine what to do about an obviously unsafe dam? I later asked Mr. Dasovich, "Are you qualified to build a dam?" "Well, I'm not a civil engineer. No, I wouldn't know how to build a dam, no."

But Mr. Dasovich did have one qualification—he knew

how to use his power to tell the sheriff what to do. Having learned that the sheriff was thinking of calling out the National Guard, and having reported in to Mr. Spotte, Mr. Dasovich now called the sheriff about 7:30 A.M. "to impress upon him that the situation was not alarming and if anyone gave him that information, it was overly done." Mr. Dasovich told him there was no need to warn the people or to call out the National Guard. As the sheriff recalls it, Mr. Dasovich "told me he had everything under control up there, that it was all right, that the dam was all right."

Thirty minutes later the dam gave way.

XXXI

Strategy

We were ready for the showdown with Pittston. We had enough facts now to prove Pittston's conduct was reckless and wanton, much more than merely negligent or careless. Once we organized these facts in the trial brief for Judge Hall, Pittston either would crack, and settle on our terms, or face a public trial and the risk of a multimillion-dollar jury verdict on punitive damages. I decided to keep a diary, so that later I could recall these events.

Tuesday, May 7: I got a call from someone who had talked with Pittston's president, Mr. Camicia, about the Buffalo Creek case. Mr. Camicia had referred to the plaintiffs and said, "They'd be happy with $10 million." He added, "Our accountant says it won't break us if we have to pay $10 million." Then, in a fairly derogatory fashion, Mr. Camicia had referred to our psychic-impairment claims. "They've got plaintiffs hundreds of miles away at the time of the flood. One preacher sued for $170,000 and he only lost $7,000. His lawyers told him he had some 'syndrome.' "

I got the impression from the way Mr. Camicia had pronounced the word "syndrome" that he did not think much of it.

Because of the way in which I found out about Mr. Camicia's $10 million figure, I worried that they might be setting a trap for us. Had they specifically arranged for us to learn about this $10 million figure? Were they telling me that they would never give more than $10 million, no matter what, and that we wouldn't even get that amount until the last possible moment before trial? If so, that meant they would continue to talk $3 million, or maybe a little more, for a long time.

Wednesday, May 8: I told Harry Huge about this conversation. He thought we were bugged, since $10 million was, in fact, the figure he would be willing to settle for. Somehow, I had to get Pittston off their $3 million offer and up to Mr. Camicia's $10 million figure as soon as possible.

I decided to let Zane know that if there was going to be a settlement it had to be completed before June 15. With the trial set for July 15, I felt I had to give myself at least a month to break down any total settlement, plaintiff by plaintiff, and get approval from each plaintiff, or fail to get it. I didn't think I'd have the time to make any settlement on the courthouse steps.

Monday, May 13: I began drafting a letter to Zane to tell him of the need to resolve the case early. I was thinking about a settlement deadline of June 1.

Tuesday, May 14: Fantastic! Judge Hall denied Pittston's motion to dismiss the psychic-impairment claims of the thirty-three plaintiffs. This was a major victory. Judge Hall had agreed with us that West Virginia law would permit recovery for mental suffering without physical impact or physical injury, if we could prove Pittston's conduct was reckless, rather than merely negligent.

Brad Butler and I had a long talk about settlement. He thought it would be "criminal for us to refuse a settlement offer of $15 million, or even $10 million." He would "even have trouble turning down an offer of $8 million." We debated how Pittston would respond to my draft letter. Brad guessed they might come back with $8 million. I guessed they might come back with $5 million.

Thursday, May 16: I talked with Bud Shay about settlement strategy. I had decided not to send a letter to Zane. Instead I would call him. I'd be calling from a position of strength, since I could say I was calling now that Judge Hall had denied their motion. Bud Shay agreed and added that I should increase our settlement proposal from $32.5 million to $35 million because of Judge Hall's ruling. We finally decided I would just say "Judge Hall's ruling supports our original settlement proposal."

Friday, May 17: I had a long telephone talk with Zane. I mentioned Judge Hall's denial of their motion. Zane interrupted and asked what motion I was talking about, since they had so many motions on file. I didn't believe Zane's feigned lack of knowledge of this particular ruling. But I went along and told him I was referring to Pittston's motion to dismiss the thirty-three absent plaintiffs. He admitted he was familiar with the fact that Judge Hall had denied that motion.

I told him I had been talking with Bud Shay about Judge Hall's decision and about our settlement discussions. I said, "Bud has suggested we raise our settlement figure now that Judge Hall has agreed with our view of the law. However, I do not intend to raise our figure at this time. But I want to be clear on where we stand on settlement." Zane interrupted and summarized matters for me, from his notes. "You made an offer of $32.5 million. We made a counteroffer of around $3 million. It was my understanding that if you thought well of it and considered it further you would be back in touch with me. I told you that I could come to Washington to visit my sick aunt and discuss settlement with you on that visit."

"Well," I said, "it's a good thing I called you, since I did not understand that the ball was in our court. You will remember that I responded to your discussion about coming to Washington for a further meeting by saying that I thought any such meeting would be premature until you and I both obtained the medical reports of all of the plaintiffs from the University of Cincinnati Medical School. Now that you have these reports I think a meeting would be appropriate." I also said, "Rightly or wrongly, I have gotten the impression that you do not fully understand the significance and magnitude of our claims for mental suffering."

Zane said, "On the contrary, I understood the significance of these claims, and the law on this subject, even before you filed your lawsuit."

I tried to tell Zane that because there were multiple plaintiffs, and because I would need to get approval of the allocations of any settlement by individual plaintiff, it would take me some time to work out a final settlement of the case. But he did not want to talk about this. He said, "I cannot recall a time when I have ever undertaken, as a defense lawyer, to involve myself in the allocation of settlement moneys among various plaintiffs." So I didn't push my argument that I needed to complete any settlement discussions by a June 1 deadline.

I did tell Zane that "the ball was now out of our court." I added, "You now have our medical reports and Judge Hall's order which we feel supports our settlement proposal. Indeed, although we feel the judge's order will support an increase in our settlement proposal, and although I had indicated to you that our $32.5 million figure probably would increase as time went on, I am still willing to discuss our original written proposal." Finally, I told him, "I am willing to meet with you in Washington or anywhere else for further settlement discussions."

Zane said, "Are you saying you are going to stay on $32.5 million?" I hoped he didn't hear me gulp when I said,

"Yes." I realized that that figure, like $64 million, was not a realistic one, but I wasn't ready to change it. He said, "I will be in touch with you."

I typed up a memo of this conversation and circulated it to a few people in our firm. The best comment I got came from Ed Brenner, one of my partners. He laughed and said, "These are wonderful negotiations—you ask for $32.5 million and they offer $3 million. Yet you say you are still negotiating, you're just $30 million apart."

Monday, May 20: Bud Vieth, Arnold & Porter's managing partner, read my memo of my conversation with Zane. He said I should have at least come down to $30 million or even $28 million. I told him I didn't want to make any new offer yet, since I was now more doubtful than ever about what our bottom-line figure should be. I told him we had just learned of a jury verdict which Don Wilson's law firm had won in a West Virginia federal court from a jury of six persons. West Virginia's federal judges had recently decided to permit six-person instead of twelve-person juries. This jury awarded $5 million damages to a plaintiff who lost money in a corporate-securities–type swindle. Of the $5 million, $3 million was punitive damages. West Virginia juries might be more willing than I had hoped to award massive punitive damages.

Bud Vieth wasn't much moved by my argument. "So settle for $25 million," he said.

Arnold Miller, the president of the United Mine Workers, was quoted in the *Charleston Gazette* that morning as saying that the plaintiffs in the Buffalo Creek case would need to win at least "$50 million" from Pittston to stop another Buffalo Creek case. This made me wonder again whether any settlement that did not air all Pittston's wrongdoing in a public trial could possibly be in the best interest of all those who had suffered for so many years at the hands of the coal companies. On the other hand, a public trial would certainly not be in the best interest of the plaintiffs, who would have to be paraded up, one by one, in front of the jury and in front of the press to tell all their psychiatric problems.

For some time now, Ken Letzler had been pushing me to meet with Jay Schulman and Dick Christie, jury experts who had received publicity for their help in picking the juries in some recent political trials, such as the Harrisburg trial and the Gainesville trial. The press had also given much credit for the acquittal of former Attorney General Mitchell and former Commerce Secretary Stans to another jury expert hired by Mitchell to provide information about the general views of people in New York. Apparently Mitchell's lawyers used this information in selecting the jurors who subsequently acquitted Mitchell and Stans.

I had tried only two jury cases, a first-degree murder case and a simple felony case. I would need all the help I could get, so Ken and I called Schulman and Christie. They were very interested in our case and agreed to meet with us in Washington in a few days.

Friday, May 24: We met with Schulman and Christie. They agreed to be our jury experts. Donovan Leisure, Pittston's New York law firm, had been using Jay Schulman as a jury expert in a case in Minnesota. Jay told us that just after we called him, Donovan Leisure asked him to work for Pittston in our case. He told them he was already working for us. This could help us. Pittston now knew we were preparing for trial, and not counting on any settlement. And we would be using jury experts they thought enough of to try and hire for their side of the case.

Wednesday, May 29: Harry Huge told me he had a dream the night before that the case was settled for $14 million.

Thursday, May 30: Zane called during the morning. "I told you I would be back to you about your statement of the possible desirability of a meeting for the purpose of settlement. We are willing to meet with you. However, since it appears that your $32.5 million is not negotiable and you do not intend to retreat from it, it appears that a meeting would be unproductive." He did not ask if in fact I would be willing to negotiate a lower figure, and instead rushed on with his monologue. "Since an overall settlement is not possible, maybe we can meet to discuss individual claims."

I told Zane, "I do not see how we can meet to discuss individual claims when the real dispute between us is the value of the psychic-impairment claims and the punitive-damage claims. The individual property claims don't seem to be of very great significance in the total picture." I then told Zane that I would be filing tomorrow a list of twelve representative cases, pursuant to Judge Hall's order. "I selected our representative cases with a view toward obtaining the jury's determination of the value of particular claims, just as you have now suggested."

Zane immediately said, "There are 600 individual lawsuits and none is alike." Obviously, Pittston would never agree that any case was representative. To do so would destroy their best weapon, delay.

I responded, "When the jury returns individual dollar figures for a few representative cases on psychic impairment and punitive damages, this should give us a range. You might not agree with the jury, and I might not agree with the jury, but still it will give us a range to talk from."

I added, "The problem appears to be that you do not value, as greatly as I do, psychic-impairment and punitive-damage claims. I have already expressed our willingness to forgo all punitive damages, but I don't think you realize the significance of that concession."

Zane laughed. "One thing I can say for you throughout this case is you have never lost your sense of humor." Zane recognized that I had not given up much when I agreed to waive any punitive damages, since our $32.5 million settlement proposal still seemed a punishing one to Pittston. "I understand you proposed no punitive damages, but you still want every last bit of damages from the compensatory portions of your cases. But it is clear that you grossly exaggerated those compensatory damage claims in your complaint. Nevertheless, we should be able to settle all the claims except as to psychic impairment." He referred to the property claims and the death cases, and said, "There is a simple formula for the death cases, $10,000, plus funeral expenses. That is the maximum that West Virginia's wrongful-death statute permits."

I told him this suggestion that we talk settlement as to property and death cases, and put aside for a moment discussion of psychic impairment, was a suggestion I was intending to make. Zane said, "Well, it appears I have read your mind."

I didn't tell Zane this, but I was becoming increasingly worried about the amount of trial time it would take just to prove property losses—cost of TVs, where did you buy the refrigerator, and so on. I did not want the jury bogged down in that kind of problem. I wanted them to focus their attention instead on our psychic-impairment claims.

We decided to meet in Charleston on June 6, the day before our scheduled pretrial conference with Judge Hall, to discuss this limited approach to settlement.

Zane then admitted, "I just wanted to be able to tell Judge Hall at the pretrial conference that we have offered to meet with the plaintiffs for settlement." Zane was trying to jockey himself into the same position I was trying to get into. So I said, "Why don't you go ahead now and make your proposal over the phone for the individual claims." He said, "I have a format for that, but I'd rather talk with you in person about it." That ended our conversation.

I filed the memorandum with Judge Hall suggesting twelve representative cases for him to choose from. Selecting our twelve was not an easy task. First, we wanted to select some plaintiffs who had been seen by two of our major medical experts—Dr. Lifton and Dr. Robert J.

Coles, the noted child psychiatrist who had seen a few of the children. However, since they had only seen a few sample plaintiffs, this limited us to a small group of plaintiffs from whom to select our representatives. Second, we truly tried to make our list as representative as possible, as Judge Hall had asked. This meant that some of our representative cases were not as dramatic as others not selected. Finally, we wanted to include some children in the list, since over 200 of our plaintiffs were children. This created an emotional problem for the parents of these children. It was hard enough to ask some of the adults to tell their story in public. It was much more difficult to ask a parent to put a child through the experience of a public trial. But each of the ten adults in our list of twelve agreed to testify in this first round of cases. And the parents of the two children also agreed, after talking with their children, to allow their children to be the first to testify about their disaster experiences.

Monday, June 3: Bud Veith asked me to explain again why I would not reduce our $32.5 million settlement proposal. I told him it wasn't time yet. I didn't want to show any weakness. Of course, there would come a time when I would have to reduce our figure, but I wanted to wait until Judge Hall pushed us before I did so.

Tuesday, June 4: I was really mad. Pittston had not filed its list of representative plaintiffs on time. This meant they had our list before them to help them decide which twelve people to list. I learned that Judge Hall was also disturbed with them for not filing their list on time.

Bud Veith called me. He finally got around to reading our memorandum to Judge Hall on the twelve representative plaintiffs. He was concerned. "That memorandum is inflammatory." I had to agree. We had quoted the findings by our doctors and Pittston's doctors on the psychiatric damages which each of the twelve people, including the two children, had suffered. Anyone reading even the admissions by Pittston's own doctor about their suffering would recognize how "inflammatory" our case was. Bud's reaction convinced me all the more that a jury would be enraged by what Pittston had done to these people.

We made a giant step forward on jury selection. Schulman and Christie were preparing a lengthy written questionnaire for pollsters to use in questioning a sample of voters from the Charleston area. The questionnaire would tell us something about which kind of juror we would like on this case—what religion, what age, what sex, what occupation, what income level, which kind of person would understand our psychiatric claims, which kind of person would be turned off by claims of mental injury, who might be willing to punish a coal company, who might not, and so forth. In other cases Schulman and Christie had used telephone polling to complete such questionnaires. They preferred live interviews with the sample voters, but that is very expensive and takes lots of time.

I noticed that the *Gazette* often carried stories about the views of the people in the Charleston area, apparently based on some kind of sample polls. I called Terry Marshall, the *Gazette* reporter who writes these stories. Terry said they had ten women on contract who conduct polls for them. They know as much about Charleston as anyone could possibly know, and Terry said they would help us. This was a godsend. We would be able to conduct careful in-person polls. With Terry's help and the help of these women, we would even be able to direct our questionnaires toward people in the very areas from which the jurors would be selected. This would give us more specific and helpful information than would questionnaires directed only to sample voters.

Thursday, June 6: Bud Shay, Harry Huge, and I met with Zane, Lauck Walton, and Dan Murdock to discuss settlement. The meeting got off to a bad start. I thought Zane was going to make an offer to us by individual plaintiff or by individual claim. Zane said he thought he was here to hear an offer from me. Lauck Walton added his two cents' worth by saying that we should end this "unseemly" argument about what we were here for. Lauck's accusing tone visibly embarrassed Zane. Harry saved the day by asking if the total amount that they might be willing to offer on an individual basis would exceed the $3 million offer they had already made. Zane said it would not. With that admission, the meeting ended. We could tell Judge Hall that they had offered only $3 million, which was so unreasonable that continuing the settlement discussion was pointless. Zane, of course, could tell Judge Hall that they were willing to talk, but we wouldn't come off our unreasonable $32.5 million proposal.

I was worried about the pretrial conference. Pittston finally filed its list of twelve cases, although they carefully insisted that no plaintiff could be representative of any other plaintiff. Their list showed that they thought Charleston jurors would not bring in large awards for black people. Of the twelve plaintiffs in Pittston's list, six were black. And three of those six were adolescent black teenagers, hardly representative of the majority of our cases. If Judge Hall used their list, we could be in trouble. I was even more concerned that Judge Hall might ask us about settlement. In a few more weeks I would be finished with our trial brief. The organization of the voluminous evidence, in understandable form in the trial brief, made Pittston's callousness seem all the more outrageous. I hoped Judge Hall would not push us to settlement until after he read it.

Friday, June 7: The pretrial conference was another victory for us. Five of the six representative plaintiffs selected by Judge Hall were from our list. The sixth, from Pittston's list, was a black woman with ten illegitimate children. That explained why Pittston selected her. Ironically, for a while we had thought of including her in our list. She was one of the best-educated plaintiffs in the entire 600, having completed two years of college. Her children were illegitimate because she did not want to get married. She had been working, taking care of ten children, and going to school until the disaster made it impossible for her to continue her schooling. Hopefully, the jury could ignore the illegitimacy of her children and her color and recognize her worth as a person.

Judge Hall also ruled with us, at least he didn't rule against us, on the numerous outstanding motions still before the court. For example, Pittston felt we still had not fully complied with all their discovery requests. And there were a number of cases which Pittston felt should be dismissed for various reasons. Judge Hall ruled that all these matters would be held in abeyance until the trial.

After the pretrial conference ended, I ran into Zane in the lobby of the Holiday Inn. He wanted to talk settlement. He apologized for the meeting the day before. He said that he was in control of the case and wanted to meet with me by himself.

I told him I too wanted to talk settlement. I said I did not want to force the plaintiffs to go through a trial. I was being very candid with him, but he immediately thought I was trying to argue our case. He said, "If you want to argue step by step, I'll argue step by step."

I told him, "I'm not arguing. I am being honest. I do not want to force the plaintiffs to trial."

I told him I also understood his problem with his co-counsel. I had seen the same thing in our firm when we were co-counsel defending a major corporate client. The jealousies between co-counsel can get quite extreme. When you get down to settlement, some lawyers spend more time trying to blame their co-counsel for losing the case than they do helping their co-counsel reach a reasonable settlement. I thought he might be having a similar problem. Finally, he said he'd be coming to Washington on business and would like to take me to lunch on Wednesday or Thursday, next week. He said he would call me.

After my talk with Zane, I began trying to figure out what my posture should be at our next settlement meeting. I tried three possibilities on Brad. My first proposal—why don't we split the difference between $32.5 million and $3 million and settle on that compromise figure, $17 million? That, in effect, would be an offer by us to settle for $17 million. My second proposal was to tell Zane that $3 million seems like a fair figure for property and wrongful death—why don't we agree on that as the total settlement figure for those claims and just have our trial on psychic impairment and punitive damages? My third proposal was to tell Zane that I had been told about the $10 million figure that Pittston thinks we would settle for. I would then say, "I reject that $10 million figure. We will never settle for $10 million." I could then add that I would be willing to come off our $32.5 million figure, but Pittston would have to come off its $10 million figure.

Brad liked the last proposal. I did too. If Zane would buy it, we would be increasing Pittston's minimum proposal from $3 million to $10 million without indicating that we were willing to settle even for $10 million. In fact, if Zane would buy it, we might be able to get even more than $10 million. Of course, Zane might say, "There is no such $10 million, but we will negotiate with you between $3 million and $10 million."

Monday, June 10: I told Bud Vieth and Ed Brenner and others at our firm about my $10 million proposal. They liked it.

Maxine Adkins, the lady who tried to call out the National Guard, told me that she would not testify for us at the trial. She was too nervous and frightened to face the ordeal of a public trial and all the anxiety that would cause her. If we subpoenaed her, she said she would get a doctor's excuse. I had to change the trial brief and delete the references to her as a witness. But I still hoped we could convince her to testify.

Wednesday, June 12: Lauck Walton called to ask for an

extension of time to file Pittston's trial brief, which was due on June 15. Lauck said Dan Murdock was ill, and he needed more time to catch up on the work Dan had been doing on the trial brief. I agreed to give them more time and wondered whether they also needed additional time for Zane to read the trial brief, since Zane always indicated to me that he read all Pittston's documents before they were filed. Incredibly, Lauck said it would be "unprecedented" if Zane read it, since "Mr. Staker hasn't read any of Pittston's filings before."

I mentioned to Lauck that I intended to meet with Zane in Washington to talk settlement. Lauck seemed surprised, and said nothing about the meeting. Afterward, it occurred to me that Zane might not have told Lauck Walton and Dan Murdock that he was coming to talk settlement with me. I decided to use this information in my meeting with Zane.

Thursday, June 13: I took Zane to lunch at the Palm Restaurant, next door to our offices. I intended to be fairly nonchalant, not pushing any talk of settlement. But as we were walking past Paul Porter's table, I decided to stop and introduce Zane to our founding partner. Paul and Zane exchanged pleasantries, and then Paul said, "Well, I hope you guys can get this thing settled today." So much for my nonchalance.

Zane and I didn't talk about settlement during lunch. I told him about my background in civil rights litigation, to let him know more about where I was coming from. He once told Bud Shay that he thought I was in this case to "raise the red flag," to get the evil corporation, instead of trying to get compensation for my clients. I wanted to play upon this feeling. I wanted Zane to think that money was secondary to me, that it might in fact be impossible for me to agree to a settlement without letting the plaintiffs have their day in court. Zane is used to negotiating with lawyers who listen to the sound of money. I thought we might have him at a disadvantage if I could convince him we were not so much interested in money as we were in establishing a principle.

After lunch, I took Zane on a tour of Arnold & Porter's townhouse offices. I even showed him our complex of "Buffalo Creek offices" with maps and file cabinets all over. I wanted him to see that many people were working on this case, preparing for trial.

Finally, we settled down in my office. I started off by laying the "Lauck Walton story" on him. I told him Lauck's silence about our meeting made me think Zane might not have the authority to talk settlement with us. I was hoping to get Zane off guard, so I would have greater credibility when I later made my $10 million pitch. Right off, Zane admitted that he had not told Lauck Walton about our meeting. "I am dealing above Lauck Walton. I have the authority to settle these cases. The client will do what I say." As for Lauck, "I told those fellows to get out from under foot."

I told Zane I understood the problems he was having on his side of the case with so many different counsel, some in New York, some in Charleston. "In a case this size, it's not surprising that everyone doesn't know what everyone else is doing or saying. In fact, there's been a leak on your side of this case which I'd better tell you about. I have learned that Pittston thinks we will be happy to settle for $10 million. I don't know how Pittston came to that conclusion, but I want you to know right here and now that we will never accept $10 million as a settlement."

Zane immediately objected that he knew of no such $10 million figure. I said I wasn't surprised that he was not aware of it, but I just wanted him to know where we stood. I had him on the defensive, with the Lauck Walton story, and now with the $10 million story. I rushed on.

"Zane, I'll come off $32.5 million, and I'll come off it substantially, but you're going to have to come off $10 million, and substantially."

Zane objected again, but more meekly now. "There isn't

any such $10 million figure. We have offered $3 million, not $10 million."

I told him I understood what he was saying, but I wanted him to know our position, and again repeated the two figures, $32.5 million and $10 million. Zane said he still didn't know what number I would be willing to accept as a settlement. I told him it was in the range substantially above 10 and substantially below 32.5, but I refused to be more specific. "If I tell you a number I am willing to settle for, I will, in effect, be agreeing to settle now for that number. As a personal matter, I don't want to settle these cases. I want to try them. In fact, I feel I owe it to the plaintiffs to give them their day in court, to publicize Pittston's recklessness and to let the world know that these plaintiffs were not responsible for this disaster, that God was not responsible, that Pittston was responsible. Only you can force me to settle, by coming up with a dollar figure I cannot in good conscience refuse to take to the plaintiffs for their consideration. You haven't come up with any such figure. You've given me the luxury of moving ahead toward trial."

Zane listened carefully. Then he told me that he was on his way to New York to talk with the Pittston people and would be back to me next Wednesday with a proposal. I told him I did not want a proposal next Wednesday. "I want Mr. Camicia to read our trial brief before you come back to us."

I was only a few days away from completing the trial brief. I had been working on it day and night for almost two months. I was convinced it was a powerful and persuasive document of corporate irresponsibility. I felt that no one, not even Mr. Camicia, could fully appreciate the strength of our case until he read this document. I also wanted to be sure the trial brief was on file with the court before we settled. In that way the true story of the Buffalo Creek disaster still would be available to the public even if we did not go to trial. Of course, the story was in the depositions and in the exhibits on file with the court in Charleston, but no one had collected all this information the way we now had in our trial brief.

Zane agreed to wait until after we filed our trial brief before getting back to me on settlement. He then flew off to New York. As soon as he left I rushed home with the worst headache I had had in months. I was more nervous and tense than I had been in a long time. I was also completely exhausted from my work on the trial brief and from the settlement meeting. After throwing up, I went to bed for two days.

Monday, June 17: Bud Vieth and I talked about my meeting with Zane. Bud thought my $10 million ploy was brilliant. "You've taken their $10 million ceiling and turned it into a $10 million floor." Ron Nathan, one of the young lawyers on the case, referred to my settlement strategy as "the big con." I only hoped it would work as well for us as it did for Newman and Redford in *The Sting*.

I mailed the trial brief to Zane with copies for him to forward to Pittston's headquarters in New York.

Wednesday, June 19: Pittston had been unable to hire any jury experts like Schulman and Christie. Schulman had given them the name of an expert, but she decided not to work for them, apparently when she learned what Pittston had done and the kind of people she would have to be defending.

The district court clerk made available to Pittston and the plaintiffs the names and addresses of the panel of eighty people from whom the jury would be selected. Each side was prohibited from making any contact with these eighty. But we could obtain general information about them from various sources. Pittston immediately had its Charleston law firm begin checking on them. And we began working with Don Wilson's law firm to obtain similar information on each of the jury panel members.

Our pollsters had just about completed their sampling.

We would run the information through a computer and soon be able to tell who should be on the jury. Indeed, the questionnaire asked the sample interviewees to estimate the amount of damages they would award in a few representative cases, so we would know the dollar verdicts we might expect. There are a number of lawyers who think all this jury work is too speculative. Still, I felt this information would better prepare us to pick the jury. Both sides would have general information about each of the eighty members on the panel. But our side also would have detailed, specific computerized information on how people who live on the streets surrounding the streets where the jurors live would evaluate our cases.

Friday, June 21: Mike White wrote a story in the *Gazette* based on Pittston's trial brief. It was headed, "Negligence in Disaster Is Denied by Pittston." Pittston didn't mind publicity based on their trial brief. But they tried to dissuade Mike and others from writing stories based on our trial brief. This was evident from the "press release" issued by the new public relations firm Pittston had just hired:

> In view of the closeness of the trial date and the substantial and unusual amount of publicity which has attended prior proceedings . . . defendant believes that it would be improper and extremely prejudicial to the interests of a fair trial for either party to use the trial memorandum as a vehicle for propagandizing as what can, at this point, be no more than mere assertion.

In addition to the attempt to shut off publicity about our trial brief, Pittston's recently hired public relations firm also tried, unsuccessfully, to hire Mike White away from the *Gazette*. Pittston's public relations people even discussed the possible benefit to Pittston of offering to spend several hundred thousand dollars in the Valley to help alleviate some of the survivors' pressing problems. I viewed these last-minute moves as desperate efforts to improve Pittston's image with the Charleston people who might sit on the jury.

Monday, June 24: There were just three weeks left till the trial. We were working feverishly to get ready. I had collected every movie I could find of the disaster, from ABC-TV, from documentary photographers, and so on. I tried to select the most gruesome, and yet most understandable, footage so the jury would understand the totality of this disaster. Harry and Ron were going around the Valley trying to line up our witnesses, going over their stories with them. They spent hours with each of the representative plaintiffs, preparing them for the trauma of the upcoming trial. They also had to rehearse the Cincinnati doctors who examined the six representative plaintiffs. And Harry was still trying to convince Maxine Adkins to come and testify for us.

We would have to spend some more time with Dr. Lifton and Dr. Erikson and Dr. Coles, and the pastoral counselors. Ken Letzler was working with Mr. Davies and Mr. Fuquay.

Roz Cohen, another member of our team of seven lawyers, had just returned from her meetings with Clinton Kennedy and Mrs. Bertsie Stanley. They would testify about Pittson's Lick Fork dam failure twenty years ago. Roz had also prepared the *Logan Banner* photographer who photographed the damage done by the failure of Dam 1 on Middle Fork in 1967. And Phil Nowak had been reviewing with Rex McGinnis his testimony about the damage to his property when Pittston's Dola, West Virginia, dam failed in 1967.

Phil was also helping me get copies of the documents we would be introducing as evidence. Ken arranged for us to use the massive scale model of the dams on Middle Fork which the U.S. Army Corps of Engineers built for the United States Senate's hearings two years earlier. We might have to rent a trailer to get that to Charleston for the trial.

Things were hectic.

I took time out to have lunch with my wife. I had just returned from lunch when Zane called. He was in New York, meeting with Pittston's people.

He said, "I waited to get back to you until we had a chance to read your trial brief. We've read it, but there was nothing surprising in it." I wasn't convinced, but held my tongue. He then said, "I've checked all around about this $10 million figure and nobody here has ever heard of any such thing."

Finally, he said, "I told Pittston your position, but they don't understand what you mean about coming off $32.5 million substantially if Pittston comes off of $10 million substantially. I told Pittston that the best I could understand was that Stern will entertain some offer between those figures." He asked if I'd be willing to come to New York to talk to him. I said, "I'll be glad to. I'll fly up tonight and stay at Arnold & Porter's New York apartment."

When I hung up I knew we had won. We were over $10 million. Anything we now got in excess of $10 million would be gravy. I had been afraid that Zane would say, "I'm willing to talk, but you should understand that under no circumstances will Pittston ever offer $10 million." If he had said that, I would have been in a box. I guess I would have refused to go and talk with him, but I'm sure the pressure on me to continue the settlement talks would have been severe.

I called Bud Shay right away and told him what had happened. I said that my position would be that I would not settle for anything less than $20 million. Now that I had $10 million I wanted to try for $15 million. To get them to 15 from 10 I probably would have to ask for 20. Zane called back before I left for New York and suggested we have dinner together. He asked me to pick a place, since he did not feel at home in New York. I suggested we try the Palm Restaurant in New York, since he had liked our lunch at their restaurant in Washington.

I rushed home, quickly got some clothes, and flew to New York. I got to Arnold & Porter's apartment earlier than I expected and instead of calling Zane to firm up the time for dinner, I decided to play it cool. I lay down on the bed and tried to rest. Zane called before many minutes passed. Obviously he was as eager as I, if not more eager, to get on with the settlement talks.

I got to the Palm Restaurant before Zane and found a table for two hidden away in a little cubbyhole called "Hernando's Hideaway." The meal was superb. We each had huge lobsters. Zane paid. The bill must have come to more than $50 because he had to give the waiter two $50 bills. We spent very little time talking about the case.

Finally, near the end of the meal Zane asked me what number I would accept to settle the case. I evaded him, saying that I personally did not want to settle the case, and that I did not want to name a number. Finally, Zane started throwing numbers at me, "What about $11 million?" I tried to stop him, to let him know that I did not want to speculate on any such numbers. He said, "$12 million? $13 million? $14 million?" I finally said, "Look, Zane, there's no sense in your going ahead this way. I'll tell you what would be reasonable. I feel that my senior partners will not let me try this case if you offer $20 million. So, if you want to take the case away from me, you can do so if you offer $20 million. At that point I lose my freedom with my firm. They will probably force me to take it." That was about it for our dinner meeting.

Tuesday, June 25: Zane called before noon, as he had promised the night before. "I reported on our meeting to the Pittston people. They are shocked at your $20 million figure. So I told them I wanted authorization to enter into final settlement discussions with you. I should have that authority by two-thirty." I told him I would wait at the apartment for his call.

Before Zane called me back, I called a friend in Portland, Maine. I wanted to know why Pittston seemed so eager to settle. Maybe Pittston's attempt to get its refinery in Maine might have something to do with this rush to settle. All I learned was that Pittston was pushing ahead with its efforts to obtain the refinery. I decided the trial date was my major lever with Pittston. The refinery, although important, was not my big chip.

Zane called back at three-thirty. "I have communicated with the people in authority. I told them I want the authority to communicate a firm and unequivocal settlement position to you as opposing counsel. This has taken a number of phone calls, one lengthy, and one with a few questions. I hope to have this firm and fast authority to come and talk with you sometime before five. I can call you in Washington before the day is over if you want to go back home." I said I'd stay.

Harry Huge called while I was waiting for Zane. I told him I thought Zane would come back with an offer of $13 million. I told Harry I would turn it down. I called Bud Vieth in Washington and reported matters to him as well.

Finally, Zane came to the apartment, with the authority to consummate a deal. While we were talking, the phone rang. It was Brad Butler wanting to know what was going on. I told Brad I'd have to talk with him later because I was meeting with Zane. A few minutes later, Zane finally made his offer, $13 million. Just as I told Zane I was not surprised by this offer, the phone rang again. It was Harry. With Zane listening, I told Harry, "I'm going to give the phone to Zane Grey Staker, and I want you to tell Zane exactly what figure I said that Pittston would offer." Harry told Zane that I had said Zane would offer $13 million. Zane said, "Well, Stern sure is prescient."

I told Zane I did not think the firm would make me accept this offer. Nevertheless, I said I should at least discuss it with the firm, especially since this was only the second time that Pittston had made a firm offer to us.

Zane and I left together for the airport. Zane was visibly happy to be leaving New York. Apparently his meetings with Pittston and Pittston's New York counsel had been difficult.

When I got to Washington I talked again with Harry and Bud Vieth. Harry laughed about his conversation with Zane. Harry said we should settle, that we had won and won big; $13 million was much more than we ever expected. Bud also said we should accept $13 million, but he gave me complete freedom to turn it down if I wanted to.

Wednesday, June 26: Everyone I talked to thought $13 million was an incredible victory. I too thought it was a good settlement for us. We would be obtaining now, from settlement, about what the pollsters' questionnaires indicated we might recover only after a lengthy trial later. On the other hand, there was no way to be certain. If we went to trial, we might do better. But there comes a point when the decision has to be made whether to proceed further or settle. I was ready to settle for $13 million, but I wanted to make one final try for more.

I called Zane after lunch. I told him the firm would not push me to settle for any figure under $15 million. I said I would split the difference with him, between $13 million and $15 million, and settle for $14 million. He immediately offered $13.5 million and then went on to talk about a number of other matters. In particular, he said that any settlement would have to include an agreement by Arnold & Porter not to represent any other plaintiffs against Pittston. I always assumed Zane would get around to asking for this. It is fairly standard practice in a major plaintiff's litigation for defense counsel to require, as part of the settlement, that the plaintiff's counsel not represent any other plaintiffs.

After we talked for a few more minutes, I finally told Zane I would accept $13.5 million.

Epilogue

The plaintiffs were overjoyed with the $13.5 million settlement. The division of the money, which we proposed and they accepted, provided them with full replacement value for their homes and possessions. This was more than their market value, the usual measure of recovery in West Virginia for these items. We also evaluated the wrongful-death cases at more than the $10,000 standard Pittston often used. This division of the $13.5 million provided approximately $5.5 million for property and wrongful-death damage payments, with approximately $8 million for the psychic-impairment, "mere puff and blow," claims. Thus, the 600 or so plaintiffs who sued for their mental injuries recovered an average of about $13,000 each.

The *net* recoveries for the plaintiffs were somewhat less, after deducting the expenses and legal fees of the case, but even then the plaintiffs recovered almost $5.5 million for their psychic-impairment claims. The 226 children alone collected a net figure of $2 million for their mental suffering as survivors. The money for the children was placed in a trust fund, with each to receive his or her portion upon reaching eighteen.

The lawsuit was a success for Arnold & Porter, too. For our time and effort, in excess of 40,000 man-hours, we earned a legal fee of almost $3 million. Sometimes you do well by doing good.

And we did some good. We made the coal company pay, and pay well. Maybe the cost of our settlement will make them a little more careful in the future. And we proved that people acting together can have some effect. They can make the legal system work for them. As plaintiff Ora Hagood put it, "The act of God was when the people banded together for a right and just cause through the process of law."

But it was a bittersweet victory. Doris Mullins, another plaintiff, explained why. "The money can help us live an easier life, free from some of our problems, but it can never put our minds completely at ease, because nothing but death can stop our minds from going back to that morning." Thus, Pug Mitchem, one of the leaders of the Buffalo Creek Citizens Committee, summed it up for ABC-TV News, "What was here is gone and will never be again."

Some Yale Law School students recently wrote a seminar paper about this case. They wondered why Pittston settled, instead of insisting that all 600 or so cases go to trial. Forcing the plaintiffs to await the outcome of all these trials might have compelled them to agree to a lower settlement figure. Ora Hagood understood this. "I don't think we could have survived a long, arduous court trial."

So the law students asked Pittston's counsel why they settled. Lauck Walton told them, "Stern was out to represent all plaintiffs everywhere who might get hurt by dirty, filthy coal companies. Most of our meetings began and ended with accusations of murder. Perhaps his attitude wore us down and speeded a settlement."

I don't think I ever used the word "murder."

Afterword

When the federal court approved our settlement it became clear to the families who had not joined our lawsuit that we had allocated more than $2 million to children for their mental suffering as survivors. Pittston had not focused on the fact that the two year statute of limitations on bringing a lawsuit did not begin to run for children until they turned eighteen. That meant hundreds of other children who survived the Buffalo Creek disaster, but had not joined our lawsuit, still could sue Pittston for their mental suffering. We immediately received numerous phone calls from families asking if we could sue for their children. We passed all those calls over to another Washington, D.C., law firm, and eventually those suits were also settled.

We also decided to help all of the survivors by creating the Buffalo Creek Foundation, funded by a $350,000 contribution from the partners of Arnold & Porter. This money financed a series of lawsuits, over the vociferous objections of the coal companies and their lawyers, to give the people of Buffalo

Creek the right to vote to incorporate themselves as a city. The coal companies fought bitterly to prevent this because an incorporated city would have the power to tax the coal companies, which owned most of the land. With that tax money, the people of Buffalo Creek could create their own police and fire departments and get better control of their lives and their community, without having to rely on the Logan County officials preferred by the coal companies.

With the help of the court system, the people finally won the right to vote on incorporation. However, the coal companies financed a disinformation campaign to make the people believe their taxes, rather than the coal companies' taxes, would go up, and the incorporation was defeated. So the Buffalo Creek Foundation paid for a police patrol car, for fire engines and a fire department built right behind Charlie Cowan's gas station, and for a health clinic for the people of Buffalo Creek.

For the first few years after we settled the Buffalo Creek litigation, I kept going back to the valley to visit with Charlie and some of the other friends I had made among our clients. My last trip to Buffalo Creek was four years ago. Charlie had died just a few years earlier, but I got to see his wife, Emma, who was packing up to move to California and live with one of her sons. She proudly showed me the beautiful house she and Charlie had built, still just up the road from Charlie's newly remodeled service station. She also took me out back to see the gleaming, polished red fire trucks in the fire department building behind their house. We talked about the difficult times she and Charlie went through when Charlie decided to bring lawyers in from Washington. She reminded me that Charlie was the quiet, calm center of the community, acting often as a buffer between the impatient and hurt people of Buffalo Creek and their lawyers, seemingly taking forever to get their cases resolved. She added, "My Charles may not have ever told you, but he thought the world of you." Nothing could have meant more to me than those words.

I also met Roland Staten at a diner for the first time since we settled his case over twenty-five years earlier. When he first walked in I didn't recognize the compact man with the sandy hair going gray. He was no longer the frightened, twenty-six-year-old coal miner I had first met. And he said he hardly recognized me without the thick, dark beard I had when I first went to Buffalo Creek. Whenever I talk about Buffalo Creek, I invariably choke up when I get to Roland's wife's last words to him, "Take care of my baby," and then Roland telling me that somewhere in that dark water he "lost that boy of mine." Roland said he had finally decided, years later, to read the Buffalo Creek disaster book after people kept telling him his story was in there. So he went to the library in Man, West Virginia, and asked for a copy. When the librarians, who had read the book, realized who he was, Roland said they broke into tears.

Roland seemed serene, maybe stoic, to me. He had remarried and though he said he had for a while talked with people about the day of the disaster, he finally decided many years ago not to give any more interviews to the press. Watching Roland as he left the diner, I realized that one of the best things we did for the survivors of the Buffalo Creek disaster was to get them to continue to talk—to us and as part of their lawsuit—about what had happened to them that day and ever since. Of course, the money damages they collected were important, but several follow-up studies of the survivors also show how important talking about what happened was to moving forward.

One of the most enduring legacies of Buffalo Creek was the decision by the American Psychiatric Association to establish a diagnosis, in the association's *Diagnostic and Statistical Manual of Mental Disorders*, for the mental suffering of survivors of a disaster like Buffalo Creek. What we called "survivor syndrome" or "psychic impairment" is now called post-traumatic stress disorder, PTSD, and its official recognition as a mental health disorder has enabled survivors of disasters like the Oklahoma City Federal Building bombing, as well as soldiers returning from war zones, to receive the mental health treatment they need to get on with their lives. From Buffalo Creek, we now know that it is critical that the survivors, friends, and family all recognize that the nightmares, the tremors, the sleeplessness, frightening flashbacks, emotional detachment, irritability, numb feelings, alcoholism, are all symptoms of what is now commonly recognized as post-traumatic stress disorder. Indeed, when 9/11 occurred, the *Wall Street Journal* sent a reporter back to Buffalo Creek to help explain to the country the post-traumatic stress New Yorkers now would be going through. His front page article in the *Wall Street Journal* was headlined "How a Tiny Stream May Help New York Recover from Tragedy—When Buffalo Creek Flooded 30 Years Ago, It Helped Define Post-Traumatic Stress."

There have been several efforts to make a movie of the Buffalo Creek disaster, but I always insisted it had to be true to the facts and not fictionalized. Then when NBC television created a docudrama unit, I gave them permission to make the Buffalo Creek disaster their first in-house produced feature movie. However, on the eve of filming, Pittston got access to a copy of the script and wrote NBC's lawyers objecting to the portrayals of Pittston and Mr. Staker and asked for equal time to respond to the movie under the "fairness doctrine" then in effect. The Federal Communications Commission, during those years, held that TV stations were "public trustees," and as such had an obligation to afford reasonable opportunity for discussion of contrasting points of view on controversial issues of public importance. Pittston was about to stand trial in the children's follow-on case to our Buffalo Creek case, so Pittston also argued the movie would prejudice their right to pick an unbiased jury in Charleston, West Virginia. NBC backed down, cancelled the Buffalo Creek movie, and completely disbanded its new docudrama unit.

The Buffalo Creek disaster book lived on though, in large

part because it became a staple for law schools to assign to incoming law students, to give them an idea of what a lawsuit is all about, not just the trials we see on television, but the investigations, legal procedural moves, and discovery fights that go on long before a trial ever occurs. To date it has sold more than a quarter million copies, and I continue to be invited to speak at law schools where the book also is used in civil procedure classes and legal writing classes. When the book was first published, the review in the *Sunday New York Times* called it a "shocking, timely book" about "crimes against ecology" that "could have been averted if Pittston . . . had been as concerned with safety as with profits." The reviewer thought the book was "particularly significant" because another major man-made disaster had just occurred at a dam in Idaho, and "the book may be an omen of the scores of similar legal and political confrontations that are bound to occur over the next decade." Well, not surprisingly, there have continued to be more and greater man-made disasters because many decision-makers at major companies still seek short-term profits at the cost of long-term safety.

Buffalo Creek led to my representing the widows of the next big coal mine disaster, the Scotia Coal Mine explosion in eastern Kentucky, which killed fifteen coal miners, nine instantly, the other six when their self-rescuers, which could provide each miner with about an hour of oxygen, gave out. One of the women had read my book and she and some of the other women asked me to come to Kentucky and represent them. At first I was reluctant, because Kentucky had a workmen's compensation statute that immunized their husbands' employer, the Scotia Coal Company, from being sued. Instead the only remedy for the miners' widows and children was a statutorily determined compensation, woefully insufficient, under the Kentucky workmen's compensation law.

When I met with the women, I learned their husbands' employer, the Scotia Coal Company, was a wholly owned subsidiary of the Blue Diamond Coal Company. I thought this might give the widows a loophole around the immunity granted Scotia. So we sued Blue Diamond, arguing that since Scotia and its parent were separate companies, Blue Diamond could not pierce its own subsidiary corporation's corporate veil and benefit from Scotia's immunity, which was opposite of what we had argued in the Buffalo Creek case.

This case took over six years to resolve. We had to overcome a hostile eastern Kentucky federal judge who dismissed our case after we had presented our evidence to a jury. After a successful appeal, we moved to recuse the hostile judge and a new, fair judge was appointed to retry our case. On the eve of the retrial, we won a substantial settlement victory for the Scotia widows and their children. But that is another story, summarized best by the *Philadelphia Inquirer*, under the heading "Fighting the Giant and Winning Big" with a picture of the Scotia widows marching determinedly to court in Pikeville, Kentucky, for the first day of their trial.

Recently, another coal mine disaster, this time in Utah, riveted many of us to our television sets, with the all-too familiar scene of grieving wives, children, and parents anxiously awaiting news about six coal miners trapped more than three miles underground from the mine entrance. We soon learned that the men had self-contained, self-rescuers, and we prayed the hastily gathered volunteer rescue teams of other coal miners could reach the men in time. Some of us recognized that our anxiety over the fate of these men was somehow connected to our own fear of being buried alive. Efforts were made to drill bore holes from the surface deep into the mine at a point where the men might have barricaded themselves to await rescue, but despite numerous bore holes in the first ten days, no signs of life could be heard or detected. Meanwhile, the rescue teams were digging feverishly through the rubble of the passageways in the underground mine, trying to reach the men. Then on the eleventh day of their herculean efforts, the mine collapsed again, killing three of the rescuers. After that, further rescue attempts were suspended, and the six men were never found.

The year before, a similar mine explosion in Sago, West Virginia, trapped thirteen coal miners two miles underground for nearly two days. By the time the rescue teams reached them, only one man was left alive, barely. This one miraculous survivor later told how four of the self-rescuers did not work, so the men shared the working units. They hammered for hours on bolts and steel plates to signal their location, but heard no response from the surface. Finally, the men began to accept their fate, prayed a little longer, and some wrote letters to their loved ones. Then they drifted off, poisoned by the carbon monoxide from the explosion.

These recent coal mine explosions show we have learned very little since Buffalo Creek or the later big coal mine explosion at the Scotia Mine in eastern Kentucky. Coal production and profits still trump safety, and the Mine Safety and Health Administration still responds more quickly to the coal operators and their lobbyists than it does to the safety needs of coal miners. We had new coal mine legislation and safety requirements after Buffalo Creek, and again after Scotia, and now again after these recent explosions in Utah, Kentucky, and West Virginia. But legislation is not enough.

Real change must come from the coal operators. They have to understand that when they cut corners on safety to get more coal out quicker, they run the risk of a disaster that will close down their mine and eliminate its profits. That happened at Buffalo Creek and Scotia, and again more recently at the Crandall Canyon Mine in Utah where the owner said, after his mine was sealed, "I'll never go near that mountain again." When coal mine operators really believe it is cheaper to operate safely, even if that cuts into production and profits in the short run, then we can expect underground coal mining to truly become safer. And if the decision-makers in other industries learn that often it is cheaper in the long run to pay for safety now at the expense of production now, those companies too will operate not only more safely, but also more profitably. The motto for those pushing safety has to be, "Pay now, or pay much more later."

—Gerald M. Stern
May 2008

Acknowledgments

Literally hundreds of people devoted their time, energy, and emotions to this lawsuit on behalf of the people of the Buffalo Creek Valley. There were the lawyers at Arnold & Porter who spent almost full time on the case, Brad Butler, Ken Letzler, Phil Nowak, Ron Nathan, and Roz Cohen, as well as Bud Shay of Steptoe & Johnson in Clarksburg, West Virginia. Other lawyers at Arnold & Porter also contributed, such as Barbara Kilberg, Ed Brenner, Bud Vieth, and Paul Porter.

There were legal assistants who spent substantially all of their time working on the lawsuit—Sally Spencer, Diane Lohman, Judy Williams, Judy Taurman, Phyllis Johnson, and Grace Venable, to name a few. And the law clerks—Stuart Madden, Craig McCabe, Roger Cohen, Charles Collins, Jim Dunlap, Mike Black, Dave Gilliatt, and Marvin Wexler, Walter B. Stuart IV, Mike Cooper, and Philip Halpern, and numerous other law clerks who worked on the case from time to time. Rickie Omohundro and Mario Magno ran our special Buffalo Creek office at Arnold & Porter. Numerous secretaries, including Jane Magged, Dottie Burd, Dyan Morgan, Carol Alliger, and Ann Ford. The Xerox operators who worked day and night at times, duplicating the mountains of paper which are the trailings of any lawsuit, and especially this one—Stan Morel and Frank Bonbrest. And there were the Arnold & Porter messengers who rushed back and forth between Washington and West Virginia and between Washington and New York with the various legal documents.

A special thanks is due to my secretaries: Stacey Carter, who typed every word of this book, often and on more than one occasion cheered me on; and Elaine Maddox, whose intelligence and friendship were shared by me and many of the survivors during the difficult early days of the case.

For the people of the Valley, I'd like to thank the experts who made their lawsuit a success: William E. Davies, Garth Fuquay, Dr. Robert Jay Lifton, Eric Olson, Dr. Kai T. Erikson, Dr. James L. Titchener, to name a few. Thanks also to the staff of the Logan-Mingo Mental Health Clinic. And finally, thanks to Judge K. K. Hall and his law clerk, Stanley Dadisman, two outstanding, sympathetic, courageous public servants.

Much credit for the book itself belongs to my agent, Bob Lescher, and my editor, Bob Loomis; to my friends Sy Hersh and Ben Wattenberg, whom I envy as writers; to my friend Leslie Schaffer, who encouraged me to continue; and to my sisters, Margot Stern Strom and Paula Stern, who told me I was great.

I am particularly grateful to my partners, Harry Huge and Mitch Rogovin—to Harry for bringing me this case and believing in my ability to handle it; to Mitch for helping me bring this story to the public.

Finally, I must acknowledge again the courage and determination of the people of the Buffalo Creek Valley, who stood up for themselves and won a new sense of their dignity and self-worth. I cannot name them all, but I also cannot fail to mention Charlie and Emma Cowan, Pug and Mable Mitchem, Roland Staten and Dennis Prince. All the rest of you, my friends in the Valley, please forgive me for not listing your names here. You are not forgotten, nor shall your struggle and triumph be forgotten.